THE THEORY OF TOPOLOGICAL SEMIGROUPS

MONOGRAPHS AND TEXTBOOKS IN PURE AND APPLIED MATHEMATICS

1. *K. Yano,* Integral Formulas in Riemannian Geometry (1970) *(out of print)*
2. *S. Kobayashi,* Hyperbolic Manifolds and Holomorphic Mappings (1970) *(out of print)*
3. *V. S. Vladimirov,* Equations of Mathematical Physics (A. Jeffrey, editor; A. Littlewood, translator) (1970) *(out of print)*
4. *B. N. Pshenichnyi,* Necessary Conditions for an Extremum (L. Neustadt, translation editor; K. Makowski, translator) (1971)
5. *L. Narici, E. Beckenstein, and G. Bachman,* Functional Analysis and Valuation Theory (1971)
6. *D. S. Passman,* Infinite Group Rings (1971)
7. *L. Dornhoff,* Group Representation Theory (in two parts). Part A: Ordinary Representation Theory. Part B: Modular Representation Theory (1971, 1972)
8. *W. Boothby and G. L. Weiss (eds.),* Symmetric Spaces: Short Courses Presented at Washington University (1972)
9. *Y. Matsushima,* Differentiable Manifolds (E. T. Kobayashi, translator) (1972)
10. *L. E. Ward, Jr.,* Topology: An Outline for a First Course (1972) *(out of print)*
11. *A. Babakhanian,* Cohomological Methods in Group Theory (1972)
12. *R. Gilmer,* Multiplicative Ideal Theory (1972)
13. *J. Yeh,* Stochastic Processes and the Wiener Integral (1973) *(out of print)*
14. *J. Barros-Neto,* Introduction to the Theory of Distributions (1973) *(out of print)*
15. *R. Larsen,* Functional Analysis: An Introduction (1973) *(out of print)*
16. *K. Yano and S. Ishihara,* Tangent and Cotangent Bundles: Differential Geometry (1973) *(out of print)*
17. *C. Procesi,* Rings with Polynomial Identities (1973)
18. *R. Hermann,* Geometry, Physics, and Systems (1973)
19. *N. R. Wallach,* Harmonic Analysis on Homogeneous Spaces (1973) *(out of print)*
20. *J. Dieudonné,* Introduction to the Theory of Formal Groups (1973)
21. *I. Vaisman,* Cohomology and Differential Forms (1973)
22. *B.-Y. Chen,* Geometry of Submanifolds (1973)
23. *M. Marcus,* Finite Dimensional Multilinear Algebra (in two parts) (1973, 1975)
24. *R. Larsen,* Banach Algebras: An Introduction (1973)
25. *R. O. Kujala and A. L. Vitter (eds.),* Value Distribution Theory: Part A; Part B. Deficit and Bezout Estimates by Wilhelm Stoll (1973)
26. *K. B. Stolarsky,* Algebraic Numbers and Diophantine Approximation (1974)
27. *A. R. Magid,* The Separable Galois Theory of Commutative Rings (1974)
28. *B. R. McDonald,* Finite Rings with Identity (1974)
29. *J. Satake,* Linear Algebra (S. Koh, T. Akiba, and S. Ihara, translators) (1975)

30. *J. S. Golan,* Localization of Noncommutative Rings (1975)
31. *G. Klambauer,* Mathematical Analysis (1975)
32. *M. K. Agoston,* Algebraic Topology: A First Course (1976)
33. *K. R. Goodearl,* Ring Theory: Nonsingular Rings and Modules (1976)
34. *L. E. Mansfield,* Linear Algebra with Geometric Applications: Selected Topics (1976
35. *N. J. Pullman,* Matrix Theory and Its Applications (1976)
36. *B. R. McDonald,* Geometric Algebra Over Local Rings (1976)
37. *C. W. Groetsch,* Generalized Inverses of Linear Operators: Representation and Approximation (1977)
38. *J. E. Kuczkowski and J. L. Gersting,* Abstract Algebra: A First Look (1977)
39. *C. O. Christenson and W. L. Voxman,* Aspects of Topology (1977)
40. *M. Nagata,* Field Theory (1977)
41. *R. L. Long,* Algebraic Number Theory (1977)
42. *W. F. Pfeffer,* Integrals and Measures (1977)
43. *R. L. Wheeden and A. Zygmund,* Measure and Integral: An Introduction to Real Analysis (1977)
44. *J. H. Curtiss,* Introduction to Functions of a Complex Variable (1978)
45. *K. Hrbacek and T. Jech,* Introduction to Set Theory (1978) *(out of print)*
46. *W. S. Massey,* Homology and Cohomology Theory (1978)
47. *M. Marcus,* Introduction to Modern Algebra (1978)
48. *E. C. Young,* Vector and Tensor Analysis (1978)
49. *S. B. Nadler, Jr.,* Hyperspaces of Sets (1978)
50. *S. K. Sehgal,* Topics in Group Rings (1978)
51. *A. C. M. van Rooij,* Non-Archimedean Functional Analysis (1978)
52. *L. Corwin and R. Szczarba,* Calculus in Vector Spaces (1979)
53. *C. Sadosky,* Interpolation of Operators and Singular Integrals: An Introduction to Harmonic Analysis (1979)
54. *J. Cronin,* Differential Equations: Introduction and Quantitative Theory (1980)
55. *C. W. Groetsch,* Elements of Applicable Functional Analysis (1980)
56. *I. Vaisman,* Foundations of Three-Dimensional Euclidean Geometry (1980)
57. *H. I. Freedman,* Deterministic Mathematical Models in Population Ecology (1980)
58. *S. B. Chae,* Lebesgue Integration (1980)
59. *C. S. Rees, S. M. Shah, and Č. V. Stanojević,* Theory and Applications of Fourier Analysis (1981)
60. *L. Nachbin,* Introduction to Functional Analysis: Banach Spaces and Differential Calculus (R. M. Aron, translator) (1981)
61. *G. Orzech and M. Orzech,* Plane Algebraic Curves: An Introduction Via Valuations (1981)
62. *R. Johnsonbaugh and W. E. Pfaffenberger,* Foundations of Mathematical Analysis (1981)

63. *W. L. Voxman and R. H. Goetschel,* Advanced Calculus: An Introduction to Modern Analysis (1981)
64. *L. J. Corwin and R. H. Szczarba,* Multivariable Calculus (1982)
65. *V. I. Istrătescu,* Introduction to Linear Operator Theory (1981)
66. *R. D. Järvinen,* Finite and Infinite Dimensional Linear Spaces: A Comparative Study in Algebraic and Analytic Settings (1981)
67. *J. K. Beem and P. E. Ehrlich,* Global Lorentzian Geometry (1981)
68. *D. L. Armacost,* The Structure of Locally Compact Abelian Groups (1981)
69. *J. W. Brewer and M. K. Smith, eds.,* Emmy Noether: A Tribute to Her Life and Work (1981)
70. *K. H. Kim,* Boolean Matrix Theory and Applications (1982)
71. *T. W. Wieting,* The Mathematical Theory of Chromatic Plane Ornaments (1982)
72. *D. B. Gauld,* Differential Topology: An Introduction (1982)
73. *R. L. Faber,* Foundations of Euclidean and Non-Euclidean Geometry (1983)
74. *M. Carmeli,* Statistical Theory and Random Matrices (1983)
75. *J. H. Carruth, J. A. Hildebrant, and R. J. Koch,* The Theory of Topological Semigroups (1983)

Other Volumes in Preparation

THE THEORY OF TOPOLOGICAL SEMIGROUPS

J. H. Carruth

Department of Mathematics
University of Tennessee
Knoxville, Tennessee

J. A. Hildebrant

R. J. Koch

Department of Mathematics
Louisiana State University
Baton Rouge, Louisiana

MARCEL DEKKER, INC. New York and Basel

Library of Congress Cataloging in Publication Data

Carruth, James Harvey
The theory of topological semigroups.

Bibliography: p. 221.
Includes index.
1. Topological semigroups. I. Hildebrant, J. A.
(John A.,). II. Koch, R. J.
III. Title.
QA387.C36 1983 512'.55 82-23560
ISBN 0-8247-1795-3

MARCEL DEKKER, INC.
270 Madison Avenue, New York, New York 10016

Current printing (last digit):
10 9 8 7 6 5 4 3 2 1

PRINTED IN THE UNITED STATES OF AMERICA

PREFACE

Our goal is to survey the field of topological semigroups, and to make this body of information accessible to the average graduate student. As a step in this direction, we offer the present volume, which lays the groundwork and covers several special avenues of recent research. The spirit of our exposition follows that of A. D. Wallace, through whose influence both in his research and by way of his students, the subject has developed since 1953. At that time his invited address to the American Mathematical Society stimulated much activity along the lines of the question "What topological spaces admit a continuous associative multiplication with unit?" As noted by Wallace, the answers to these questions are likely to involve more algebra and topology than was the case for compact groups, where there is a representation theory due to the presence of Haar measure. Our coverage here includes background material, internal structure, products, quotients, and semigroups with some special algebraic or topological property. We defer to a subsequent volume some of the aspects which rely on cohomology or category theory.

This material can serve as a text for an introduction to topological semigroups. It is not intended as a research tract, but rather to expose various aspects of the subject.

In the initial chapter we discuss the fundamental concepts of topological semigroups, their substructures, and maps between semigroups. Most of the notions here were introduced in the 1955 Wallace notes, and these will be further explored in the remaining chapters. Featured in the final section of this chapter is the Lawson-Madison theorem which generalizes an earlier result of Wallace on compact semigroups to k_ω-semigroups. We present their proof for locally compact σ-compact semigroups.

In the second chapter we demonstrate various techniques of developing new examples of semigroups from existing ones, and discuss the properties of these new semigroups inherited from the old semigroups. Constructions include free topological semigroups, Bohr compactifications, and products of various types.

Monothetic semigroups, which were characterized by the work of Koch, Numakura, and Hewitt, are discussed in the first section of the third chapter. The Wallace-Rees-Suschkewitch structure theorem for compact completely simple semigroups appears in this chapter. Most of the remaining portions of the chapter are devoted to algebraic considerations of Green's relations and quasi-orders.

Contributions of Clifford, Faucett, Mostert and Shields, Cohen and Krule, and Phillips are featured in the fourth chapter on interval semigroups (threads). Their structure and the nature of congruences on interval semigroups are characterized in this brief chapter.

The fifth chapter features the Carruth-Lawson proof of the classical Mostert-Shields theorem on the existence of one parameter semigroups in certain compact monoids. An important example due to Hunter appears at the end of the chapter.

In the final chapter we present results on compact divisible semigroups. This chapter features the results of Keimel, the contributions of Hudson and Hofmann, the structure theorems of Brown and Friedberg, and the Hildebrant characterization of compact subunithetic semigroups (the atoms of compact divisible semigroups).

More than a reference source for the text, the bibliography of this book is one of the most complete guides to the literature in topological semigroups to date, listing nearly 400 articles and books.

J. H. Carruth
J. A. Hildebrant
R. J. Koch

CONTENTS

Chapter 1

FUNDAMENTAL CONCEPTS

In this chapter we present some concepts which we consider to be fundamental to the study of topological semigroups. Many of the results pertain to compact semigroups, as will be the situation for the entire book. However, some of the theorems are purely algebraic in nature or apply to a more general class of topological semigroups. The algebraic results are included to make the chapter as self-contained as we consider feasible. Some knowledge of elementary topology will be assumed. The notation in this chapter will be used throughout the book.

SEMIGROUPS

A *semigroup* is a non-empty set S together with an associative multiplication $(x,y) \to xy$ from $S \times S$ into S. The associative condition on S states that $x(yz) = (xy)z$ for each $x, y, z \in S$. If A and B are subsets of S, we use the notation $AB = \{ab : a \in A \text{ and } b \in B\}$. If S has a Hausdorff topology such that $(x,y) \mapsto xy$ is continuous, with the product topology on $S \times S$, then S is called a *topological semigroup*. The condition that the multiplication on S is continuous is equivalent to the condition that for each $x, y \in S$ and each open set W in S with $xy \in W$, there exist open sets U and V in S such that $x \in U$, $y \in V$, and $UV \subset W$.

If the word "semigroup" appears with a topological adjective, then "topological semigroup" is implied. For example, the statement "S is a compact semigroup" means that S is a compact topological semigroup. Observe that any semigroup can be made into a topological semigroup by giving it the discrete topology, and thus a finite semigroup is a compact semigroup.

Relations are one of the fundamental notions in the study of topological semigroups. We recall some of the basic concepts pertaining to relations. A *relation on a set* X is a subset of $X \times X$. The *diagonal* of X is the relation $\Delta(X) = \{(x,x) : x \in X\}$. When no confusion seems likely, we write simply Δ for the diagonal relation. If A and B are relations on X, then the *composition* of A with B is $A \circ B = \{(x,y) : (x,z) \in A \text{ and } (z,y) \in B \text{ for some } z \in X\}$. The *converse* of a relation R on X is $R^{-1} = \{(x,y) \in X \times X : (y,x) \in R\}$. A relation R on X is *reflexive* if $\Delta \subset R$, *symmetric* if $R = R^{-1}$, *transitive* if $R \circ R \subset R$, and an *equivalence* if R is reflexive, symmetric, and transitive. A relation R on X is *anti-symmetric* if $R \cap R^{-1} \subset \Delta$, a *quasi-order* if R is reflexive and transitive, and a *partial order* if R is an anti-symmetric quasi-order. If R is a relation on a set X, then $R^{(n)}$ is defined by $R^{(1)} = R$ and $R^{(n+1)} = R^{(n)} \circ R$ for each positive integer n. The *transitive closure* of R is defined $\mathrm{Tr}(R) = \cup \{R^{(n)} : n \text{ a positive integer}\}$. It is readily seen that $\mathrm{Tr}(R)$ is the smallest transitive relation of X containing R.

Although nets are not an essential part of the study of topological semigroups, they sometimes serve as a convenient tool in establishing results about topological semigroups. We recall a few basic notions about nets and directed sets.

A *directed set* is a pair $(D,\leqslant)$, where D is a non-empty set and $\leqslant$ is a reflexive and transitive relation on D such that for $\alpha, \beta \in D$ there exists $\gamma \in D$ such that $\alpha \leqslant \gamma$ and $\beta \leqslant \gamma$. Notice that we indicate that $(\alpha,\gamma) \in \leqslant$ in the traditional manner by $\alpha \leqslant \gamma$. A subset C of D is *cofinal* in D if for each $\alpha \in D$, there exists $\beta \in C$ such that $\alpha \leqslant \beta$, and a subset R of D is *residual* in D if there exists $\delta \in D$ such that if $\delta \leqslant \gamma$ in D, then $\gamma \in R$. If no confusion seems likely, we suppress the mention of $\leqslant$ and write simply D for the directed set.

A *net* in a set X is a function from a directed set D into X. We denote the image of $\alpha \in D$ by x_α and the net itself by $\{x_\alpha\}_{\alpha\in D}$ or simply $\{x_\alpha\}$ when no confusion seems likely. If T is a subset of X, then the net $\{x_\alpha\}_{\alpha \in D}$ is *frequently* in T provided $\{\alpha \in D : x_\alpha \in T\}$ is cofinal in D, and $\{x_\alpha\}_{\alpha \in D}$ is *eventually* in T if $\{\alpha \in D : x_\alpha \in T\}$ is residual in D. If X is a space, $x \in X$, and $\{x_\alpha\}$ is a net in X, then $\{x_\alpha\}$ *clusters* to x if $\{x_\alpha\}$ is frequently in each neighborhood of x, and $\{x_\alpha\}$ *converges* to x if $\{x_\alpha\}$ is eventually in each neighborhood of x. In the latter case we write $\{x_\alpha\} \to x$. A *subnet* of a net $f : (D,\leqslant) \to X$ is a pair (g,ψ), where $g : (E,\prec) \to X$ is a net and $\psi : E \to D$ is a function such that $g = f \circ \psi$ and for each $\alpha \in D$, there exists $\beta \in E$ such that if $\beta \prec \gamma$ in E, then $\alpha \leqslant \psi(\gamma)$ in D.

There are some useful topological results pertaining to nets. A net in a Hausdorff space converges to at most one point; A space X is compact if and only if each net in X has a convergent subnet; A subset C of a space X is closed if and only if for each net $\{x_\alpha\}$ in C converging to a point $x \in X$, we have $x \in C$; and a function f from a space X into a space Y is continuous if and only if $\{x_\alpha\} \to x$ in x implies $\{f(x_\alpha)\} \to f(x)$ in Y. Convergence can be replaced by clustering in Y or by clustering in both X and Y in the last result.

If S is a Hausdorff space endowed with an associative multiplication, then continuity of multiplication on S is equivalent to the statement that for nets in S, $\{x_\alpha\} \to x$ and $\{y_\alpha\} \to y$ implies that $\{x_\alpha y_\alpha\} \to xy$. Convergence of the net $\{x_\alpha y_\alpha\}$ can be replaced by clustering if one of the nets $\{x_\alpha\} \to x$ or $\{y_\alpha\} \to y$ is assumed only to cluster to the given point.

The following is one of the more useful topological results in the area of topological semigroups [Wallace, 1951]:

<u>*1.1 Theorem*</u>. Let X, Y, and Z be spaces, A a compact subset of X, B a compact subset of Y, $f : X \times Y \to Z$ a continuous function, and W an open subset of Z containing $f(A \times B)$. Then there exists an open set U in X and an open set V in Y such that $A \subset U$, $B \subset V$, and $f(U \times V) \subset W$.

Proof. Since f is continuous, $f^{-1}(W)$ is an open set in $X \times Y$ containing $A \times B$. For each (x,y) in $A \times B$, there exist open sets M and N in X and Y, respectively, such that $x \in M$, $y \in N$, and $M \times N \subset f^{-1}(W)$. Since B is compact, for a fixed $x \in A$, there are open sets

$M_1,\ldots,M_n$ in X containing x and corresponding open sets $N_1,\ldots,N_n$ in Y such that $B \subset Q = N_1 \cup \cdots \cup N_n$. Let $P = M_1 \cap \cdots \cap M_n$. Then P is open in X, Q is open in Y, $x \in P$, $B \subset Q$, and $P \times Q \subset f^{-1}(W)$. Since A is compact, there exists open sets $P_1,\ldots,P_m$ in X and corresponding $Q_1,\ldots,Q_m$ open in Y such that $B \subset V = Q_1 \cap \cdots \cap Q_m$ and $A \subset U = P_1 \cup \cdots \cup P_m$. It follows that U and V are the required open sets. ∎

In 1.1, if one has the additional hypothesis that X is locally compact, then U can be chosen so that $\overline{U}$ is compact, and likewise if Y is locally compact, then V can be chosen so that $\overline{V}$ is compact.

1.2 Theorem. Let A and B be subsets of a topological semigroup S.

(a) If A and B are compact, then AB is compact.

(b) If A and B are connected, then AB is connected.

Proof. This is immediate from the fact that $AB = m(A \times B)$, where $m : S \times S \to S$ is multiplication. ∎

Observe that "compact" or "connected" can be replaced in 1.2 by any topological property which is productive and preserved by continuous functions, e.g., "arcwise connected".

1.3 Theorem. Let A and B be subsets of a topological semigroup S. Then:

(a) If B is closed, then $\{x \in S : xA \subset B\}$ is closed;

(b) If B is compact, then $\{x \in S : A \subset xB\}$ is closed;

(c) If B is compact, then $\{x \in S : xA \subset Bx\}$ is closed;

(d) If A is compact and B is open, then $\{x \in S : xA \subset B\}$ is open; and

(e) If A is compact and B is closed, then $\{x \in S : xA \cap B \neq \square\}$ is closed.

Proof. To prove (a), let $C = \{x \in S : xA \subset B\}$ and fix $y \in S\backslash C$. Then there exists $a \in A$ such that $ya \in S\backslash B$. Since $S\backslash B$ is open, there exist open sets M and N in S such that $y \in M$, $a \in N$, and $MN \subset S\backslash B$. In particular, we have that $Ma \subset S\backslash B$. It follows that $y \in M \subset S\backslash C$, $S\backslash C$ is open, and hence C is closed.

To prove (b), let $F = \{x \in S : A \subset xB\}$ and fix $y \in S \backslash F$. Then $a \notin yB$ for some $a \in A$. Since yB is compact, there exist disjoint open sets M and N such that $a \in M$ and $yB \subset N$. In view of 1.1, we see that there exist open sets U and V such that $y \in U$, $B \subset V$, and $UV \subset N$. In particular, $UB \subset N$, $UB \cap M = \Box$, $a \notin UB$, $U \subset S \backslash F$, $S \backslash F$ is open, and hence F is closed.

To prove (c), we apply 1.1 and the continuity of multiplication in S to conclude that $\{x \in S : xA \subset Bx\}$ is closed if B is compact.

To prove (d), we again apply 1.1.

To prove (e), we apply (d). ■

To illustrate the usefulness of nets in topological semigroups, we present a net argument as an alternate proof of 1.3 (a).

To prove (a) using nets, let $\{x_\alpha\}$ be a net in $C = \{x \in S : xA \subset B\}$ such that $\{x_\alpha\} \to x$ with $x \in S$, and let $a \in A$. Then $\{x_\alpha a\} \to xa$ in B, and since B is closed, $xa \in B$. We obtain that $xA \subset B$, $x \in C$, and hence C is closed.

If S is a semigroup, $a \in S$, and n is a positive integer, then a^n is defined recursively by $a^1 = a$ and $a^{k+1} = a^k a$.

An element e of a semigroup S is called an *idempotent* if $e^2 = e$. The set of idempotents of S is denoted by $E(S)$, or when no confusion seems likely, simply by E.

The set of idempotents of a semigroup may be empty, as is the case for the additive semigroup of positive integers. We will show later that if S is a compact semigroup, then $E \neq \Box$. Moreover, in any topological semigroup S, we have that E is closed. The latter result we obtain as a consequence of the following topological result:

1.4 Theorem. Let X be a Hausdorff space and $f : X \to X$ a continuous function. Then the set of fixed points of f is closed in X.

Proof. Define $g : X \to X \times X$ by $g(x) = (f(x),x)$. Then g is continuous and so $g^{-1}(\Delta(X)) = \{x \in X : f(x) = x\}$ is closed, since the diagonal of a Hausdorff space X is closed in $X \times X$. ■

1.5 Theorem. If S is a topological semigroup, then E(S) is a closed subset of S.

Proof. This is immediate from 1.4 and the observation that E is the set of fixed points of the continuous function $x \to x^2$. ∎

1.6 Theorem. Let S be a topological semigroup. For e, f ∈ E(S), define e ≤ f if ef = fe = e. Then ≤ is a partial order on E and is a closed subspace of S × S.

Proof. The argument that ≤ is a partial order on E is straightforward. To see that ≤ is closed in S × S, let $\{(e_\alpha, f_\alpha)\} \to (e,f)$ in S × S with $\{(e_\alpha, f_\alpha)\}$ in ≤. Then $\{e_\alpha\} \to e$ and $\{f_\alpha\} \to f$, and since E is closed in S, we have that (e,f) ∈ E × E. Moreover, $e_\alpha f_\alpha = f_\alpha e_\alpha = e_\alpha$ for all α, so that the continuity of multiplication yields that ef = fe = f. We conclude that (e,f) ∈ ≤ and ≤ is closed. ∎

If S is a semigroup and a ∈ S, then the function x → xa is called *right translation* by a and is denoted ρ_a, and x → ax is called *left translation* by a and is denoted λ_a. It is clear that if S is a topological semigroup and a ∈ S, then both ρ_a and λ_a are continuous. Moreover, in the case that S is a topological semigroup and e ∈ E, we have that ρ_e is a retraction of S onto Se, λ_e is a retraction of S onto eS, and $\rho_e \circ \lambda_e = \lambda_e \circ \rho_e$ is a retraction of S onto eSe.

A semigroup S is said to be *abelian* (or *commutative*) if ab = ba for all a, b ∈ S.

An element e of a semigroup S is called a *left identity* for S if ex = x for all x ∈ S, a *right identity* for S if xe = x for all x ∈ S, and an *identity* for S if e is both a left and right identity. Observe that if e is either a left or right identity for S, then e ∈ E. Moreover, if S has a left identity e and a right identity f, then e = ef = f is an identity for S. Thus each semigroup S can have at most one identity.

A semigroup which has an identity is called a *monoid*.

If S is a [topological] semigroup, we can adjoin an identity 1 to S [discretely] to form a new [topological] semigroup T = S ∪ {1}. Note that if S is a compact semigroup, then T is a compact semigroup.

If S is a [topological] semigroup, then $S^1 = S$ if S has an identity, and $S \cup \{1\}$ [with 1 adjointed discretely] otherwise.

An element f of a semigroup S is called a *left zero* for S if $fx = f$ for all $x \in S$, a *right zero* for S if $xf = f$ for all $x \in S$, and a *zero* for S if f is both a left and right zero for S. As in the case of a left or right identity, a left or right zero for a semigroup S is an idempotent, and S can have at most one zero. If S is a [topological] semigroup, we can adjoin [discretely] a zero element in a manner analogous to that of adjoining an identity. We also define S^0 to be the obvious analog of S^1.

If S is a semigroup with a zero 0, and x is an element of S such that $x^n = 0$ for some positive integer n, then x is called a *nilpotent element* of S.

A *band* is a semigroup S such that $S = E(S)$, and a *semilattice* is a commutative band. Semilattices have been the topic for extensive study in both of the fields of algebraic and topological semigroups. A chapter of Volume II will be devoted to this topic.

We now turn to the presentation of a collection of examples of topological semigroups. Notation used in these examples will be applied throughout this book.

Let $\mathbb{N}$ be the additive semigroup of positive integers with the discrete topology. Then $\mathbb{N}$ is a non-compact abelian topological semigroup. Observe that $E(\mathbb{N}) = \square$. Let $\mathbb{N}^* = \mathbb{N} \cup \{\infty\}$ denote the one-point compactification of $\mathbb{N}$ with $x + \infty = \infty + x = \infty$ for each $x \in \mathbb{N}^*$. Then $\mathbb{N}^*$ is a compact abelian semigroup with $E(\mathbb{N}^*) = \{\infty\}$, and ∞ is a zero.

Let $\mathbb{H}$ denote the additive semigroup of non-negative real numbers with the usual topology. Then $\mathbb{H}$ is a locally compact σ-compact connected abelian semigroup with $E(\mathbb{H}) = \{0\}$, and 0 is an identity. Let $\mathbb{H}^* = \mathbb{H} \cup \{\infty\}$ denote the one-point compactification of $\mathbb{H}$ with $\infty + x = x + \infty = \infty$ for each $x \in \mathbb{H}^*$. Then $\mathbb{H}^*$ is a compact connected abelian semigroup with $E(\mathbb{H}^*) = \{0,\infty\}$, 0 is an identity, and ∞ is a zero.

One can convert any non-empty Hausdorff space S into a topological semigroup by declaring that $xy = x$ for all $x, y \in S$. A semigroup with this multiplication is called a *left zero semigroup*. If we define $xy = y$ for all $x, y \in S$, then S is called a *right zero semigroup*. For a left zero semigroup, the multiplication on S is simply first projection $\pi_1 : S \times S \to S$, and second projection if S is a right zero semigroup.

If S is a Hausdorff space, $z \in S$, and we define $xy = z$ for all $x, y \in S$. Then S is called a *zero semigroup* , and is an abelian topological semigroup with zero z and $E = \{z\}$.

There are three fundamental types of semigroups on an interval which, as we will see later, are the building blocks of what we will call I-semigroups. We present these three basic examples now.

Let $I_u = [0,1]$ be the real unit interval with the usual topology and usual multiplication. Then I_u is a compact abelian semigroup called the *usual interval*. Note that 0 is a zero, 1 is an identity, $E = \{0,1\}$, and I_u has no nilpotent elements except 0. We shall show later that I_u and $\mathbb{H}^*$ are the same semigroup in a certain sense.

Let $I_n = [1/2,1]$ with the usual topology and multiplication $(x,y) \mapsto \min\{x,y\}$. Then I_m is a compact semilattice, 0 is a zero, and 1 is an identity. The semigroup I_m is called the *min interval*.

Let $I_n = \{1/2,1\}$ with the usual topology and multiplication $(x,y) \to \max\{1/2,xy\}$, where xy is the usual product of x and y. Then I_n is a compact abelian semigroup, $E = \{1/2,1\}$, 1/2 is a zero, 1 is an identity, and each element of $I_n \backslash \{1\}$ is nilpotent. The semigroup I_n is called the *nilpotent interval*.

SUBSEMIGROUPS

A considerable portion of the study of topological semigroups deals with determining the algebraic and topological structure of subsemigroups of a given class of topological semigroups. For example, each compact connected monoid contains an irreducible subsemigroup, and one of the major results in compact semigroup theory is that irreducible semigroups are abelian [Hofmann and Mostert, 1966]. In this section we develop the concept of a subsemigroup and give some examples.

If A is a subset of a semigroup S and $n \in \mathbb{N}$, then A^n is defined recursively by $A^1 = A$ and $A^{k+1} = A^k A$.

A *subsemigroup* of a semigroup S is a non-empty subset T of S such that $T^2 \subset T$.

Observe that a subsemigroup T of a [topological] semigroup S is itself a [topological] semigroup under the restriction of the multiplication on S to $T \times T$.

If A is a subset of a semigroup S, then the set T of all finite products of elements of A is the smallest subsemigroup of S containing A and T is called the *subsemigroup of* S *generated by* A. Observe that $T = \cup \{A^n : n \in \mathbb{N}\}$.

It is evident that the intersection (if non-empty) of a collection of subsemigroups of a semigroup S is again a subsemigroup of S. To see that the union of subsemigroups need not be a subsemigroup, consider the semigroup $\mathbb{N}$ and the semigroups $\{2n : n \in \mathbb{N}\}$ and $\{3n : n \in \mathbb{N}\}$.

Certain subsemigroups of a given semigroup appear with such regularity in the remainder of this book that they are isolated here and some of their basic properties are presented.

If S is a semigroup and $x \in S$, then $\theta(x) = \{x^n : n \in \mathbb{N}\}$ is an abelian subsemigroup of S and is the subsemigroup of S generated by x. If $A \subset S$, then $Z(A) = \{b \in S : ba = ab \text{ for all } a \in A\}$ is called the *centralizer* of A in S, and $N(A) = \{b \in S : bA = Ab\}$ is called the *normalizer* of A in S. The set Z(S) is called the *center* of S. If $N(A) = S$, then A is said to be *normal* in S. If $N(S) = S$, then S is called a *normal semigroup*.

If S is a topological semigroup and A and B are subsets of S, observe that $\overline{A}\,\overline{B} \subset \overline{AB}$. If $\overline{A}$ and $\overline{B}$ are both compact then $\overline{A}\,\overline{B} = \overline{AB}$ since in this case, $\overline{A}\,\overline{B}$ is closed and contains AB.

From the preceding observations one can conclude that if T is a subsemigroup of a topological semigroup S, then $\overline{T}$ is also a subsemigroup of S. Moreover, if T is abelian, then so is $\overline{T}$. To see this, let $f : S \times S \to S \times S$ be defined by $f(x,y) = (xy,yx)$. Then f is continuous and hence $f^{-1}(\Delta)$ is closed. Now if T is abelian, then

$T \times T \subset f^{-1}(\Delta)$, and since $\overline{T} \times \overline{T} = \overline{T \times T} \subset f^{-1}(\Delta)$, we have that $\overline{T}$ is abelian.

If A is a subset of a topological semigroup S, then the smallest closed subsemigroup of S containing A is the closure of the subsemigroup of S generated by A. In particular, if $x \in S$, then $\Gamma(x) = \overline{\theta(x)}$ is a closed subsemigroup of S called the *monothetic subsemigroup of* S *with generator* x. If $S = \Gamma(x)$, then S is called a *monothetic semigroup*.

1.7 Theorem. Let S be a topological semigroup, $x \in S$, A a non-empty subset of S, and $e \in E(S)$. Then:

(a) $\theta(x)$ is an abelian subsemigroup of S and is the smallest subsemigroup of S containing x;

(b) $\Gamma(x)$ is an abelian subsemigroup of S and is the smallest closed subsemigroup of S containing x;

(c) eS is a closed subsemigroup of S with left identity e;

(d) Se is a closed subsemigroup of S with right identity e;

(e) $eSe = eS \cap Se = \{x \in S : ex = x = xe\}$ is a closed subsemigroup of S with identity e;

(f) Z(A) is a closed subsemigroup of S if $Z(A) \neq \Box$;

(g) $Z(A) \subset N(A)$;

(h) N(A) is a subsemigroup of S if $N(A) \neq \Box$, and is closed if A is compact;

(i) AS and SA are subsemigroup of S, and if S and A are compact, then AS and SA are closed; and

(j) if $E(S) \subset N(S)$, then $E(S) \subset Z(S)$. ∎

We close this section by presenting some examples of semigroups and mentioning certain of their distinguished subsemigroups.

Let $\mathbb{C}$ denote the space of complex numbers with complex multiplication. Then $\mathbb{C}$ is an abelian topological semigroup with a zero and an identity and no other idempotents. The non-zero elements of $\mathbb{C}$ form a non-closed subsemigroup as do the non-zero real elements of

$\mathbb{C}$ and the positive real elements of $\mathbb{C}$. The real elements of $\mathbb{C}$ form a closed subsemigroup as do the non-negative real elements of $\mathbb{C}$. The unit disk $\mathbb{D} = \{z \in \mathbb{C} : |z| \leqslant 1\}$ is a compact subsemigroup as are the circle $\{z \in \mathbb{D} : |z| = 1\}$, the interval $\{z \in \mathbb{C} : z$ is real and $|z| \leqslant 1\}$, and the interval $\{z \in \mathbb{C} : z$ is real and $0 \leqslant z \leqslant 1\}$. If $z \in \mathbb{C}$, then $\Gamma(z) = \theta(z)$ if $|z| \geqslant 1$, $\Gamma(z) = \theta(z) \cup \{0\}$ if $|z| < 1$, and $\Gamma(z) \subset \{z \in \mathbb{C} : |z| = 1\}$ if $|z| = 1$. Let $n \in \mathbb{N}$ and let A_n denote the nth roots of unity (solutions to the equation $z^n = 1$). Then B_n, the convex hull of A_n in $\mathbb{C}$, is a compact subsemigroup of $\mathbb{D}$.

For $n \in \mathbb{N}$, let $M(n,\mathbb{C})$ denote the set of all $n \times n$ matrices with complex entries under ordinary matrix multiplication. Then $M(n,\mathbb{C})$ is a semigroup and is non-abelian if $n > 1$. If $M(n,\mathbb{C})$ is given the topology obtained by identifying it with $\mathbb{C}^{n^2}$, then $M(n,\mathbb{C})$ becomes a topological semigroup. The subset $Gl(n,\mathbb{C})$ of non-singular matrices is a subsemigroup of $M(n,\mathbb{C})$ as is the set of diagonal matrices and the set of lower triangular matrices. One can form the semigroups $M(n,\mathbb{R})$ and $Gl(n,\mathbb{R})$, where $\mathbb{R}$ is the space of real numbers, in the same manner.

If $S = M(2,\mathbb{C})$ and $e = \begin{bmatrix} 1 & 0 \\ 0 & 0 \end{bmatrix}$, then $e \in E(S)$, $eS = \{\begin{bmatrix} x & y \\ 0 & 0 \end{bmatrix} : x, y \in \mathbb{C}\}$,

$$Se = \{\begin{bmatrix} x & 0 \\ y & 0 \end{bmatrix} : x, y \in \mathbb{C}\}$$

$$eSe = \{\begin{bmatrix} x & 0 \\ 0 & 0 \end{bmatrix} : x \in \mathbb{C}\}, \quad \text{and}$$

$$N(\{e\}) = Z(\{e\}) = \{\begin{bmatrix} x & 0 \\ 0 & y \end{bmatrix} : x, y \in \mathbb{C}\}$$

Let $T = \{\begin{bmatrix} x & y \\ 0 & 1 \end{bmatrix} \in M(2,\mathbb{R}) : 0 \leqslant x \leqslant 1,\ 0 \leqslant y \leqslant 1, \text{ and } x + y \leqslant 1\}$. Then T is a compact subsemigroup of $M(2,\mathbb{R})$ called the *triangle*. The underlying space of T is homeomorphic to the triangle in the first quadrant bounded by $x + y = 1$, the x-axis, and the y-axis. Each bounding interval is a closed subsemigroup as is each line segment joining a point $(0,x)$ to $(1,0)$. For $(x,y) \in T$ with $x > 0$, the subsemigroup $(x,y)T$ is illustrated in the following figure:

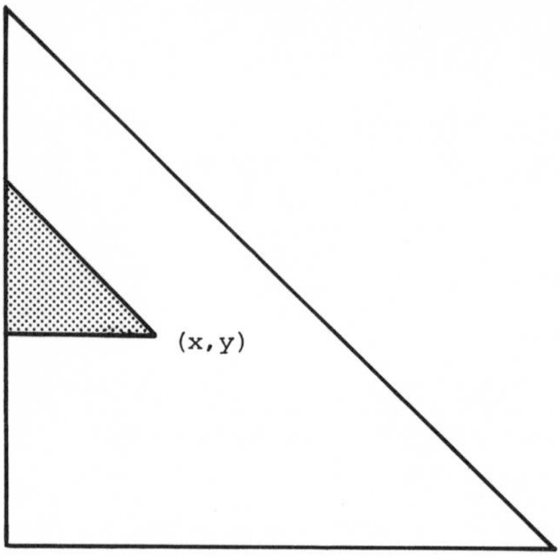

and the subsemigroup $T(x,y)$ is illustrated in the following figure:

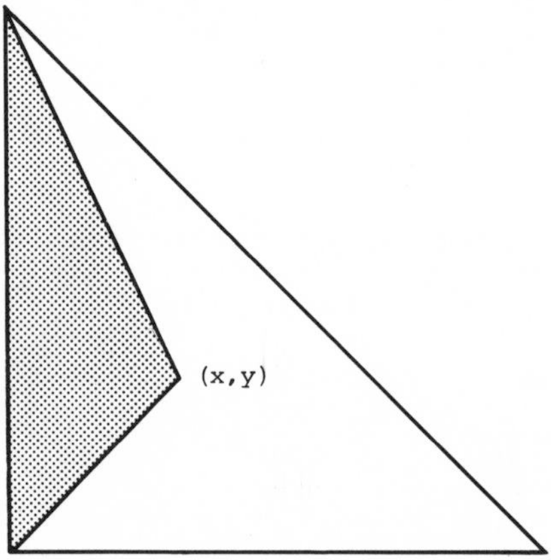

The element $(1,0)$ is an identity for T and $E(T) = \{(1,0)\} \cup \{(0,x) : 0 \leqslant x \leqslant 1\}$ is a subsemigroup. Observe that $N(T) = Z(T) = \{(1,0)\}$.

Let X be a set and let $S = X^X$ be the set of all functions from X into X under composition. Then S is a semigroup and is non-abelian if X contains more than one element. The set of all injections is a subsemigroup of S as is the set of all surjections. Hence, the set of bijections is a subsemigroup of S. If X is finite, then S is finite, and hence compact.

SUBGROUPS

In this section we turn our attention to subgroups. Along with numerous basic results most of which appear in [Wallace, 1955], we present a proof of the fact that each compact semigroup contains an idempotent [Numakura, 1952], an interesting new proof of the Swelling Lemma [Wallace, 1953], and a proof that each locally compact semigroup which is algebraically a group is a topological group [Ellis, 1957a]. It will become evident that subgroups play an essential role in the theory of topological semigroups.

A *subgroup* of a semigroup S is a non-empty subset G of S such that $xG = G = Gx$ for each $x \in G$. Observe that a subgroup of a semigroup is algebraically a group and is a subsemigroup.

Recall that a family $\mathcal{F}$ of subsets of a set X is a *descending family* (or *filter base*) if for each $A, B \in \mathcal{F}$, there exists $C \in \mathcal{F}$ such that $C \subset A \cap B$. If for each $A, B \in \mathcal{F}$, we have that either $A \subset B$ or $B \subset A$, then $\mathcal{F}$ is called a *tower*.

If X is a compact space, $\mathcal{F}$ is a descending family of non-empty closed subsets of X, and U is an open set in X with $\cap \mathcal{F} \subset U$, then there exists $F \in \mathcal{F}$ such that $F \subset U$. In particular, $\cap \mathcal{F} \neq \square$. To see this, observe that $X\backslash U \subset X \cap \mathcal{F} = \cup\{X\backslash F : F \in \mathcal{F}\}$ and so $X\backslash U \subset (X\backslash F_1) \cup \cdots \cup (X\backslash F_n) = X\backslash(F_1 \cap \cdots \cap F_n)$ for some finite subcollection. $F_1, \ldots, F_n$ of $\mathcal{F}$. But, as $\mathcal{F}$ is descending, there exists $F \in \mathcal{F}$ such that $F \subset F_1 \cap \cdots \cap F_n$, and it follows that $F \subset U$. We employ this result in the proof of the next result.

1.8 Theorem. A compact semigroup S contains an idempotent.

Proof. We show that S contains a minimal closed subsemigroup and that every such subsemigroup consists of a single idempotent. Hence, the idempotents of S are precisely the minimal closed subsemigroups of S. Let $\mathcal{S}$ denote the set of closed subsemigroups of S. Note that $S \in \mathcal{S}$, and hence $\mathcal{S} \neq \square$. Partially order $\mathcal{S}$ by inclusion and let $\mathcal{C}$ be a maximal tower in $\mathcal{S}$, by use of the Hausdorff Maximality Principal (see [Kelly, 1955]). Let $T = \cap \mathcal{C}$. Then, as we observed in the previous paragraph, $T \neq \Box$. Let $x \in T$. Then xT is a closed subsemigroup of S contained in T. In view of the maximality of $\mathcal{C}$, we see that $xT = T$, and similarly $Tx = T$. Thus T is a subgroup of S. If e is the identity of T, the maximality of $\mathcal{C}$ ensures that $T = \{e\}$. ∎

It is a straightforward matter to construct continuous non-associative multiplications on compact spaces with no idempotents. Such multiplications exist on any non-degenerate finite space. Hence, associativity plays an essential role in 1.8. We mention, however, that if a space X has the fixed point property, then any continuous multiplication on X has an idempotent, since $x \to x^2$ has a fixed point. Many of the results contained in this book do not really depend on associativity of multiplication.

Next we prove a remarkably useful theorem about translates of compact sets in compact semigroups [Wallace, 1953].

1.9 The Swelling Lemma. Let S be a compact semigroup and let A be a closed subset of S. If $t \in S$, and $A \subset tA$, then $A = tA$.

Proof. Let $T = \{x \in S : tA \subset xA\}$. Then for $x, y \in T$, we have $tA \subset xA \subset xtA \subset xyA$, so that T is a subsemigroup of S. In view of 1.3(b), we see that T is closed, and hence by 1.8 contains an idempotent e. We obtain that $A \subset tA \subset eA$. For fixed $a \in A$, there exists $b \in A$ such that $a = eb$. Hence, $ea = e(eb) = eb = a$ and it follows that $eA \subset A$. Therefore, $eA = A$ and from $A \subset tA \subset eA$ we obtain the desired result that $A = tA$. ∎

The proof of 1.9 can be modified to yield the result that if S is a topological semigroup, A is a compact subset of S, and t is an element of some compact subsemigroup of S such that $A \subset tA$, then $A = tA$.

One of the problems that has given impetus to the theory of topological semigroups is the problem of finding topological and/or algebraic hypothesis on a semigroup which imply that it must be a group (see [Wallace, 1955]). The next theorem is one of the earliest results of this type, [Iwassawa, 1948] (see also [Gelbaum, Kalisch, and Olmsted, 1951]).

A semigroup S is said to be *left cancellative* provided $x, y, z \in S$ and $xy = xz$ imply that $y = z$. If $x, y, z \in S$ and $yx = zx$ imply that $y = z$, then S is said to be *right cancellative*. If S is both left and right cancellative, then S is said to be *cancellative*.

1.10 Theorem. Let S be a compact cancellative semigroup. Then S is a group.

Proof. Let $e \in E$. Then for $x \in S$, we have $ex = e^2x = e(ex)$, so that $x = ex$ by cancellation, and similarly $x = xe$. It follows that e is an identity of S and $E = \{e\}$. Let $p \in S$. Then pS is a compact subsemigroup of S, and hence $ps \cap E \neq \square$, so that $e \in pS$. But then $S = eS \subset (pS)S \subset pS$, so that $pS = S$ by 1.9. Similarly $Sp = S$, and we conclude that S is a group. ∎

A slight modification of the proof of 1.10 yields the stronger result that if S is a cancellative topological semigroup such that each element of S lies in a compact subsemigroup of S, then S is a group.

As a direct consequence of 1.10 we have that if T is a compact subsemigroup of a topological semigroup S and S is algebraically a group, then T is a group.

A subgroup G of a topological semigroup is a *topological group* if the map $x \to x^{-1}$ sending x to its inverse is continuous on G.

If S is the semigroup of real numbers under addition with a base for the topology of S consisting of intervals closed on the left and open on the right, then S is a topological semigroup which is algebraically a group. However, S is not a topological group, since inversion $x \to -x$ is not continuous.

We now establish a result that will be employed later.

1.11 Theorem. Let S be a compact semigroup and let G be a subgroup of S. Then $\overline{G}$ is a subgroup of S.

Proof. For each $x \in G$, $xG = G$ implies $x\overline{G} = \overline{xG} = \overline{G}$. Now combining 1.3 (a) and (b), we obtain that $A = \{x \in S : x\overline{G} = \overline{G}\}$ is a closed subset of G containing G. Hence $\overline{G} \subset A$ and $x\overline{G} = \overline{G}$ for each $x \in \overline{G}$. Similarly, $\overline{G}x = \overline{G}$ for each $x \in \overline{G}$ and hence $\overline{G}$ is a subgroup of S. ∎

If S is the multiplicative semigroup of non-negative real numbers and $G = S \backslash \{0\}$, then S is a locally compact semigroup and G is a subgroup of S. However, $\overline{G} = S$ is not a group. Thus 1.11 is not generally valid for locally compact semigroups.

We turn now to an important topological result which will be employed now and later in this book. First recall that if X is a space and Y is a compact space, then first projection $\pi_1 : X \times Y \to X$ is a closed map. To see this, let C be a closed subset of $X \times Y$ and suppose $x \in X \backslash \pi_1(C)$. Then $(\{x\} \times Y) \cap C = \square$, and by 1.1, $(U \times Y) \cap C = \square$ for some open set U containing x. It follows that $U \subset X \backslash \pi_1(C)$ and $\pi_1(C)$ is closed.

1.12 Lemma. Let X be a space, Y a compact Hausdorff space, and $f : X \to Y$ a function. Then f is continuous if and only if Graph $(f) = \{(x,f(x)) : x \in X\}$ is a closed subspace of $X \times Y$.

Proof. Suppose that f is continuous, and let $g : X \times Y \to Y \times Y$ be defined by $g(x,y) = (f(x),y)$. Then g is continuous and hence Graph $(f) = g^{-1}(\Delta(Y))$ is closed.

Suppose that Graph (f) is closed, and let C be a closed subset of Y. Then $f^{-1}(C) = \pi_1((X \times C) \cap \text{Graph }(f))$ is closed, and hence f is continuous. ∎

1.13 Theorem. Let G be a compact semigroup which is a group. Then G is a topological group.

Proof. Let m be the multiplication function on G, and let e denote the identity of G. Then the graph of the inversion function on G is

$$\{(x,y) \in G \times G : m(x,y) = e\} = m^{-1}(e)$$

which is closed, since m is continuous. In view of 1.12, we see that inversion is continuous on G and hence G is a topological group. ∎

We will prove later in this section the stronger result of Ellis which establishes that 1.13 is valid if "compact" is replaced by "locally compact".

If S is a semigroup and G and H are subgroups of S such that $G \cap H \neq \Box$, then $G \cap H$ is a subgroup of S. It is not generally true that the union of groups in S is a subgroup of S. However, the union of all groups containing a given idempotent of S is a subgroup of S. We will now establish this algebraic result.

> *1.14 Theorem.* Let S be a semigroup, $e \in E$, and H(e) the union of all subgroups of S containing e. Then H(e) is a subgroup of S.

Proof. Now e acts as an identity for H(e) and hence for the subsemigroup G generated by H(e). Let $g \in G$. Then $g = g_1 g_2 \ldots g_n$, where $g_1, g_2, \ldots, g_n \in H(e)$. Observe that $g_n^{-1} \ldots g_2^{-1} g_1^{-1}$ is an inverse for g, where g_i^{-1} is an inverse of g_i in a subgroup of S containing g_i; $i = 1,2,\ldots,n$. Hence G is a group and G = H(e). ■

If S is a semigroup and $e \in E$, then the union of all subgroups of S containing e is the maximal subgroup of S containing e. We will use H(e) to denote the *maximal subgroup of* S *containing* e.

A word about the terminology seems in order at this point. In a partially ordered set $(X,\leqslant)$, $m \in X$ is maximal provided $m \leqslant x$ implies $m = x$, whereas $m \in X$ is the greatest element of X if $x \leqslant m$ for all $x \in X$. It is easy to construct an example of a partially ordered set which has a unique maximal element but no greatest element. The concepts are obviously equivalent for partially ordered sets with greatest elements. It would actually be more descriptive of the situation to define H(e) to be the greatest subgroup of S containing e. However, the maximal subgroup terminology is now standard and we adopt it here.

If S is a semigroup and $e \in E$. Then a useful fact is that $H(e) = eSe \cap \{x \in S : e \in xS \cap Sx\}$.

For a semigroup S, we let $H(S) = \{H(e) : e \in E\}$. When no confusion seems likely, we write H for H(S). Let $\mathcal{E} : H \to H$ be defined by $\mathcal{E}(x) = e$ if $x \in H(e)$ and $e \in E$, and let $\mathcal{I} : H \to H$ be defined by $\mathcal{I}(x) = x^{-1}$, where x^{-1} is the inverse of x in $H(\mathcal{E}(x))$.

It is a straightforward argument to prove that if S is a compact semigroup, then H is closed, $\mathcal{I}$ is continuous, and $\mathcal{E}$ is a retraction of H onto E.

Neither E nor H need be a subsemigroup of a semigroup S. To see this, let S be the semigroup consisting of the real matrices $\begin{bmatrix} 0 & 0 \\ 0 & 0 \end{bmatrix}, \begin{bmatrix} 1 & 0 \\ 0 & 0 \end{bmatrix}, \begin{bmatrix} 1 & -1 \\ 0 & 0 \end{bmatrix}, \begin{bmatrix} 0 & 0 \\ -1 & 1 \end{bmatrix}$ and $\begin{bmatrix} 0 & 0 \\ -1 & 0 \end{bmatrix}$ under ordinary matrix multiplication. Then $E = H = S \backslash \begin{bmatrix} 0 & 0 \\ -1 & 0 \end{bmatrix}$, and $\begin{bmatrix} 0 & 0 \\ -1 & 0 \end{bmatrix} = \begin{bmatrix} 0 & 0 \\ -1 & 1 \end{bmatrix} \begin{bmatrix} 1 & 0 \\ 0 & 0 \end{bmatrix} \in H^2 H$.

Combining 1.13, 1.14, and our observations we obtain the following rather comprehensive result:

1.15 Theorem. Let S be a semigroup and e E. Then $H(e) = eSe \cap \{x \in S : e \in xS \cap Sx\}$ is the union of all subgroups of S containing e. If $e \neq f$ in E, then $H(e) \cap H(f) = \square$. If S is compact, then H(e) is a compact topological group for each $e \in E$, $H = \cup\{H(e) : e \in E\}$ is a closed subspace of S. The map $\mathcal{E}$ is a retraction of H onto E, and $\mathcal{I}$ is continuous (where $\mathcal{E}(x) = e$ if $x \in H(e)$ and $e \in E$, and $\mathcal{I}(x)$ is the inverse of x in $H(\mathcal{E}(x))$).

Recall that a monoid is a semigroup S with an identity 1. Of particular importance among the subgroups of S is the group H(1). We call H(1) the *group of units* of S and an element of H(1) is called a *unit*. There are three useful characterizations of H(1). These are expressed: $H(1) = \{x \in S : 1 \in xS \cap Sx\} = \{x \in S : \lambda_x \text{ is bijective}\} = \{x \in S : \rho_x \text{ is bijective}\}$.

In a more general setting we have

1.16 Theorem. Let S be a semigroup and let $e \in E$. Then H(e) is the group of units of eSe.

Proof. Now e is an identity for eSe and $H(e) = eSe \cap \{x \in S : e \in xS \cap Sx\}$. Therefore, H(e) is a subgroup of eSe containing its identity e, and maximality of H(e) in S implies maximality of H(e) in eSe. ∎

If S is a topological monoid and 1 denotes the identity of S, then we can characterize H(1) as follows: $H(1) = \{x \in S : \lambda_x$ is a homeomorphism from S onto S$\} = \{x \in S : \rho_x$ is a homeomorphism from S onto S$\}$. If S is compact, then the word "homeomorphism" can be replaced by "surjection" in this characterization.

As a matter of standard notation in a monoid S, we will use 1 to denote the identity unless otherwise specified or confusion seems likely.

1.17 Theorem. Let S be a compact monoid and let $x, y \in S$ such that $xy \in H(1)$. Then $x \in H(1)$ and $y \in H(1)$.

Proof. Now $S = Sxy \subset Sy$ implies that $S = Sy$ by 1.9. We also have that $yS \subset S = xyS = x(yS)$, so that $yS = xyS$ by 1.9; and since $xyS = S$, we have $S = yS$. Now $1 \in yS \cap Sy$ and hence $y \in H(1)$. Similarly, $x \in H(1)$. ∎

We demonstrate that 1.17 need not hold for non-compact monoids by the following example:

Let $\mathbb{B}$ be the set of ordered pairs of non-negative integers with multiplication defined by $(a,b)(c,d) = (a + c - \min\{b,c\}, b + d - \min\{b,c\})$. Then $\mathbb{B}$ is a discrete monoid, $1 = (0,0)$ is the identity for $\mathbb{B}$ and $H(1) = \{1\}$. Moreover, if $x = (0,1)$ and $y = (1,0)$, then $xy = 1 \in H(1)$ but neither x nor y lies in H(1). The monoid $\mathbb{B}$ is called the *bicyclic semigroup* and it may be "realized" as the semigroup with identity element 1 generated by a two element set $\{p,q\}$ subject to the single generating relation $qp = 1$. (See [Clifford and Preston, 1961].) We also observe that $E(\mathbb{B}) = \{(a,a) : a \geqslant 0\}$ and $H(\mathbb{B}) = E(\mathbb{B})$. (See example following 3.37.)

Recall that a space X is *homogeneous* if given any $x, y \in X$, there exists a homeomorphism f from X onto X such that $f(x) = y$.

Observe that a topological semigroup S which is algebraically a group is homogeneous, since $S = H(1)$ and for $x, y \in S$, we have that $\lambda_{yx^{-1}}$ is a homeomorphism from S onto S such that $\lambda_{yx^{-1}}(x) = y$. In particular, a topological group is homogeneous.

We next establish an important result regarding locally compact semigroups (see [Ellis, 1957b]). We will employ the notation that if A is a subset of a group, then $A^{-1} = \{a^{-1} : a \in A\}$, where a^{-1} is the inverse of a.

1.18 Theorem. Let S be a locally compact semigroup which is algebraically a group. Then S is a topological group.

Proof. We establish this result in a sequence of four steps.

Step 1. *If* A *is a compact subset of* S, *then* A^{-1} *is closed.* Let $\{x_\alpha\}$ be a net in A^{-1} with $\{x_\alpha\} \to x$ in S, and let y be a cluster point of $\{x_\alpha^{-1}\}$ in A. Then $\{1\} = \{x_\alpha x_\alpha^{-1}\}$ clusters to xy, and hence $xy = 1$. We conclude that $x = y^{-1} \in A^{-1}$ and A^{-1} is closed.

Step 2. *If* A *is a countable subset of* S, *then* $\overline{A}^{-1} \subset \overline{A^{-1}}$. Fix $x \in \overline{A}$. We will show that $x^{-1} \in \overline{A^{-1}}$. For this purpose, let $B_n = (A \cup \{x, x^{-1}\} \cup A^{-1})^n$ for each $n \in \mathbb{N}$. Then $B = \cup\{B_n : n \in \mathbb{N}\}$ is a countable subgroup of S and $\overline{B}$ is a subsemigroup of S. Let V be a compact neighborhood of 1 and let $c \in \overline{B}$. Then cV is a neighborhood of c and so $cV \cap B \neq \square$. This implies that $c \in BV^{-1}$ and hence $\overline{B} \subset BV^{-1}$. Thus $\overline{B} = \cup\{bV^{-1} \cap \overline{B} : b \in B\} = \cup\{b(V^{-1} \cap \overline{B}) : b \in B\}$. According to Step 1, V^{-1} is closed and hence $b(V^{-1} \cap \overline{B})$ is closed for each $b \in B$. Now, $\overline{B}$ is locally compact and so the interior relative to $\overline{B}$ of one of the sets $b(V^{-1} \cap \overline{B})$ is not empty according to the Baire category theorem (see p. 250 of [Dugundji, 1966]). Fix such a point $b \in B$, an open set U in S, and an element $c \in B$ such that $c \in U \cap B \subset U \cap \overline{B} \subset b(V^{-1} \cap \overline{B})$. Then, $xc^{-1}(U \cap \overline{B}) = xc^{-1}U \cap \overline{B}$ and $xc^{-1}U = W$ is open. Now $W \cap A \subset W \cap \overline{B} \subset xc^{-1}(U \cap W) \subset xc^{-1}bV^{-1}$ and so $(W \cap A)^{-1} \subset Vb^{-1}cx^{-1} = C$ with C compact, since V is compact. By Step 1, $\overline{(W \cap A)^{-1}}^{-1}$ is a closed subset of S containing $W \cap A$. Hence $\overline{W \cap A} \subset \overline{(W \cap A)^{-1}}^{-1}$ and $x^{-1} \in \overline{(W \cap A)}^{-1} \subset \overline{(W \cap A)^{-1}} \subset \overline{A^{-1}}$. We conclude that $\overline{A}^{-1} \subset \overline{A^{-1}}$.

Step 3. *If* A *is a compact subset of* S, *then* A^{-1} *is compact.* First observe that A^{-1} is closed by Step 1. Let V be a compact neighborhood of 1 and assume that A^{-1} cannot be covered by a finite number of compact sets $x_i^{-1}V$ with $x_i \in A$. Then there exists $\{x_i^{-1} : i \in \mathbb{N}\} \subset A^{-1}$ such that $x_n^{-1} \notin \cup\{x_i^{-1}V : i = 1,2,\ldots,n-1\}$ for $n \in \mathbb{N}$. Let $E_n = \{x_i : i \geqslant n\}$ and let $\{x_i\}$ cluster to $x \in A$. Let $m \in \mathbb{N}$ such that $x_m \in V$. Then $x^{-1} \in x_m^{-1}V$. Clearly, $x \in \overline{E}_{m+1}$ and so, by Step 2, $x^{-1} \in E_{m+1}^{-1}$. Hence, there exists $n > m$ such that $x_n^{-1} \in x_m^{-1}V$, contradicting the choice of $\{x_i^{-1} : i \in \mathbb{N}\}$.

Final Step. *Inversion is continuous.* Fix $x \in S$ and let U be an open set containing x^{-1}. Let F be the family of compact neighborhoods of x. Then F is a descending family such that $\cap F = \{x\}$. It follows that $F^{-1} = \{V^{-1} : V \in F\}$ is a descending family of compact subsets (by Step 3) and $\cap F^{-1} = \{x^{-1}\} \subset U$. Hence there exists $V \in F$ such that $x^{-1} \in V^{-1} \subset U$ and the result is obtained. ■

It was shown later [Ellis, 1957b] that if multiplication is only separately continuous, i.e. λ_x and ρ_x are continuous for each x, on a locally compact Hausdorff space G which is algebraically a group, then G is a topological group.

A celebrated result of E. Cartan states that if the n-sphere is a topological group, then n = 0, 1, or 3 [Cartan, 1936]. In the following example we exhibit group multiplication on the 0, 1, and 3 spheres.

Let

$$D_Q = \{\begin{bmatrix} z & w \\ -\overline{w} & \overline{z} \end{bmatrix} \in M(2,\mathbb{C}) : \det \begin{bmatrix} z & w \\ -\overline{w} & \overline{z} \end{bmatrix} \leqslant 1\},$$

where $\overline{Z}$ denotes the conjugate of Z in $\mathbb{C}$. Then D_Q, called the *quaternionic disk*, is a compact connected semigroup with identity $\begin{bmatrix} 1 & 0 \\ 0 & 1 \end{bmatrix}$ and the group of units is $T = \{x \in D_Q : \det x = 1\}$. Moreover, the underlying spaces of T is a 3-sphere. The unit circle in $\mathbb{C}$ forms a topological group on the 1-sphere, and the two element group is a topological group on the 0-sphere.

In the semigroup $\mathbb{C}$, we observe that H(1) is the set of non-zero elements of $\mathbb{C}$ and is an open subgroup of $\mathbb{C}$ such that $0 \in \overline{H(1)}$.

In M(n,$\mathbb{C}$) and M(n,$\mathbb{R}$), the groups of units are the groups of non-singular matrices.

A matrix B in M(n,$\mathbb{C}$) is called *unitary* if B^{-1} is the transpose of the matrix whose (i,j) entry is the conjugate of the (i,j) entry of B, i.e., B^{-1} is the conjugate transpose of B. The set U(n,$\mathbb{C}$) of unitary matrices forms a compact connected subgroup of M(n,$\mathbb{C}$). It is called the *unitary* n-*group*.

In the group M(n,$\mathbb{R}$) the collection O(n,$\mathbb{R}$) of orthogonal matrices (those whose inverse is the transpose) form a compact subgroup of M(n,$\mathbb{R}$) with two components (see [Chevalley, 1946]) and is called the *orthogonal* n-*group*.

In the group U(n,$\mathbb{C}$) the collection SU(n,$\mathbb{C}$) of matrices with determinant 1 forms a compact connected group called the *special unitary* n-*group*.

In the group U(n,$\mathbb{R}$) the collection SO(n,$\mathbb{C}$) of matrices with determinant 1 forms a compact connected group called the *special orthogonal* n-*group*.

The orthogonal groups are very basic in the theory of topological groups. The classical theorem of Peter and Weyl yields that every compact group is embeddable in a product of orthogonal groups [Peter and Weyl, 1927].

In the triangle semigroup, all subgroups are trivial.

Let X be a set and let $S = X^X$ under composition. Then 1_X, the identity map on X, is the identity of S and $H(1_X)$ is the group of permutations of X. Maximal groups of constant mappings are trivial.

For notational purposes we will use $\mathbb{Z}$ to denote the subgroup of $\mathbb{R}$ consisting of the set of all integers. Note that $\mathbb{Z}$ is discrete.

IDEALS

As is true in the case of ring theory, ideals play a vital role in the theory of semigroups. In this section we introduce various types of ideals and observe that all types of ideals that have been studied in semigroup theory may be considered as invariant sets of certain semigroups of transformations.

A non-empty subset [L,R] I of a semigroup S is a [*left*, *right*] *ideal* of S if [$SL \subset L$, $RS \subset R$] $IS \cup SI \subset I$.

Observe that each left (or right) ideal of a semigroup S is a subsemigroup of S, and each ideal of S is both a left and a right ideal of S. Also note that the union and non-empty intersection of a collection of [left, right] ideals is again a [left, right] ideal. Observe that if S is abelian, then the notions of left ideal, right ideal, and ideal are all the same.

If A is a non-empty subset of a semigroup S, then S itself is a [left, right] ideal of S containing A. The intersection of all [left, right] ideals containing A is evidently the smallest [left, right] ideal of S containing A. It is called the [*left*, *right*] *ideal generated by* A.

A [left, right] ideal of a semigroup S is called a *minimal* [*left*, *right*] *ideal* if it properly contains no other [left, right] ideal.

If A is a subset of a semigroup S, then $L(A) = S^1A = A \cup SA$, $R(A) = AS^1 = A \cup AS$, and $J(A) = S^1AS^1 = A \cup SA \cup AS \cup SAS$.

If A is a [left, right] ideal of a semigroup S, then A is called a *principal* [*left*, *right*] *ideal* if there exists $x \in S$ such that A is the [left, right] ideal of S generated by x. In this case we say that A is the *principal* [*left*, *right*] *ideal of* S *generated by* x.

We gather together several elementary facts about ideals in the next theorem.

1.19 Theorem. Let S be a semigroup, L a left ideal, R a right ideal, I and J ideals, G a subgroup, and A a non-empty subset of S. Then:

(a) [SA,AS] SAS is a [left, right] ideal of S;

(b) [L(A),R(A)] J(A) is the [left, right] ideal of S generated by A;

(c) $RL \subset R \cap L$, and hence $R \cap L \neq \square$;

(d) $I \cap J$ is an ideal;

(e) If $G \cap I \neq \Box$ $[G \cap L \neq \Box,\ G \cap R \neq \Box]$, then $G \subset I$ $[G \subset L,\ G \subset R]$;

(f) If $[L,R]$ J is a minimal [left, right] ideal and $[x \in L,\ x \in R]$ $x \in J$, then $[L = L(x),\ R = R(x)]$ $J = J(x)$;

(g) If I and J are minimal ideals, then $I = J$; and

(h) If I is a group, then I is a minimal ideal.

Proof. The proof of (a) is transparent.

To prove (b), let L be the left ideal generated by A. Then $A \subset L$, so that $SA \subset SL \subset L$. It follows that $A \subset L(A) \subset L$, and since $L(A)$ is clearly a left ideal, we have that $L(A) = L$. Similar arguments work for $R(A)$ and $J(A)$.

The proofs of (c) and (d) are straightforward.

To prove (e), suppose that $G \cap L \neq \Box$ and take a $p \in G \cap L$. Then $G = Gp \subset SL \subset L$. Again a similar argument works for R, and the conclusion for J follows from the conclusion for L, since J is also a left ideal.

For the remaining parts we have that (f) is a consequence of (b), (g) is a consequence of (e), and (b) is straightforward. ■

If S is a semigroup and $\cap\{I : I \text{ is an ideal of } S\}$ is non-empty empty, then we denote this intersection by $M(S)$.

In view of 1.19 (g), if a semigroup S contains a minimal ideal, then it contains exactly one and this unique minimal ideal is $M(S)$. Moreover, if S has a zero 0, then $M(S) = \{0\}$.

As we will see, a compact semigroup contains minimal ideals of all three types. In general, however, the existence of a minimal ideal need not imply the existence of a minimal left ideal or the existence of a minimal right ideal. This is demonstrated in the following example:

Consider the bicyclic semigroup $\mathbb{B}$. If $(a,b) \in \mathbb{B}$, then $(a,b)\mathbb{B} = \{(a,n) : n \geqslant 0\}$, $\mathbb{B}(a,b) = \{(n,b) : n \geqslant 0\}$, and $\mathbb{B}(a,b)\mathbb{B} = \mathbb{B}$. The last equality yields immediately that $\mathbb{B} = M(\mathbb{B})$. However, if L is any left ideal and $(a,b) \in L$, then $\mathbb{B}(a,b) = \{(n,b) : n \geqslant 0\} \subset L$, $L' = \mathbb{B}(a,b+1) = \{(n,b+1) : n \geqslant 0\} \subset L$, and L'

is a left ideal of $\mathbb{B}$ which is properly contained in L. Therefore, $\mathbb{B}$ has no minimal left ideals and similarly $\mathbb{B}$ has no minimal right ideals. Hence, $\mathbb{B}$ is an example of a semigroup which contains a minimal ideal but which contains neither minimal left nor minimal right ideals.

The existence of a minimal ideal and a minimal left ideal in a semigroup need not imply the existence of a minimal right ideal. This is demonstrated in the following example:

Let S be the subsemigroup of $\mathbb{N}^{\mathbb{N}}$ consisting of those injective functions f such that $\mathbb{N}\backslash f(\mathbb{N})$ is infinite. Then for $f \in S$, $Sf = S$ and so S is itself a minimal left ideal. It follows also that $M(S) = S$. However, $f \notin fS$ as follows: If $g \in S$ and $p \in \mathbb{N}\backslash g(\mathbb{N})$, then $g(p) \neq p$ and so $f(g(p)) \neq f(p)$ (thus $fg \neq f$). Thus, if R is a right ideal of S and $f \in R$, then $R' = fS$ is a right ideal of S properly contained in R. Therefore, S is a semigroup which contains a minimal ideal, a minimal left ideal, but no minimal right ideal. This semigroup is a particular example of a broad class of semigroups called Baer-Levi semigroups (see section 8.1 of [Clifford and Preston, 1961]).

Those semigroups having group minimal ideals are frequently of interest. The next theorem gives a sufficient condition for a semigroup S with a minimal ideal to have that M(S) is a group.

1.20 Theorem. If S is a semigroup containing a minimal ideal such that $M(S) \subset Z(S)$, then M(S) is a group.

Proof. Let $x \in M(S)$. Then for $y \in S$, $yxM(S) = xyM(S) \subset xM(S)$ and $xM(S)y \subset xM(S)$, so that xM(S) is an ideal of S. In view of the minimality of M(S), we see that $xM(S) = M(S)$, and similarly $M(S)x = M(S)$. ■

1.21 Corollary. If S is an abelian semigroup containing a minimal ideal, then M(S) is a group.

1.22 Theorem. If S is a semigroup containing a minimal ideal, M(S) is a group, and e is the identity of M(S), then $e \in Z(S)$.

Proof. Fix $x \in S$. Then ex and xe are in $M(S)$ and so $ex = exe = xe$. ■

The next Theorem appears in [Koch, 1953].

1.23 Theorem. Let S be a semigroup and let $e \in E(S)$. Then the following are equivalent:

(a) Se is a minimal left ideal;

(b) eSe is a group;

(c) eS is a minimal right ideal.

Moreover, if S is compact, then each of (a), (b), and (c) is equivalent to:

(d) SeS is the minimal ideal of S.

Proof. To see that (a) implies (b), fix $x \in eSe$. Then $Sx \subset Se$ and so $Sx = Se$ in view of the minimality of Se. Hence, $e \in Sx$ and there is a $y \in S$ such that $e = yx$. It follows that $e = eyex$ and so x has a left inverse in eSe. Therefore, eSe is a group [van der Waerden, 1949].

For the proof that (b) implies (c), let R be a right ideal of S with $R \subset eS$. Then, if a R we have $ae \in Re \subset eSe$ and so $e \in ae(eSe) \subset aS$. Therefore, $eS \subset aS \subset R$ and eS is a minimal right ideal.

The proofs that (c) implies (b) and (b) implies (a) are similar to those above.

In case S is compact we show that (c) and (d) are equivalent. Suppose then that S is compact.

For (c) implies (d), suppose eS is a minimal right ideal. Clearly SeS is an ideal. If I is an ideal contained in SeS, then $eIS \subset eS$ and so $eIS = eS$ by the minimality of eS. We obtain $SeIS = SeS \subset I$ (since I is an ideal). Thus SeS is the minimal ideal of S.

For (d) implies (c), suppose SeS is the minimal ideal of S. Let R be a right ideal of S contained in eS and fix $x \in R$. Then $xS \subset R \subset eS$ and $SxS \subset SeS$. Therefore, $SxS = SeS$ by minimality. Hence, $e = axb$ for some $a, b \in S$. Now $xS \subset R \subset eS = axbS \subset axS$. By the Swelling Lemma, we have $xS = R = eS = axS$, and hence eS is a minimal right ideal. ■

1.24 Theorem. Let [L,R] I be a [left, right] ideal of a topological semigroup S. Then $[\overline{L},\overline{R}]$ $\overline{I}$ is a [left, right] ideal of S.

Proof. Now $S\overline{L} = \overline{S}\,\overline{L} \subset \overline{SL} \subset \overline{L}$, and hence $\overline{L}$ is a left ideal of S. Similar arguments work for R and I. ■

If A is a non-empty subset of a topological semigroup S, then the smallest closed [left, right] ideal of S containing A is evidently the intersection of all closed [left, right] ideals of S containing A, and is called the closed [left, right] ideal of S generated by A. In view of 1.24, we see that $[\overline{L(A)},\overline{R(A)}]$ $\overline{J(A)}$ is the closed [left, right] ideal of S generated by A.

1.25 Theorem. Let S be a compact [connected] semigroup [with identity] and let A be a compact [connected] subset of S. Then L(A), R(A), and J(A) are compact [connected].

Proof. In the case that A and S are compact, we have $L(A) = A \cup SA$, and SA is compact. It follows that L(A) is compact. In the case that S is connected and has an identity, and A is connected, we have $L(A) = A \cup SA = SA$. It follows that L(A) is connected. Similar arguments work for R(A) and J(A) in each case. ■

1.26 Corollary. Let S be a compact [connected] semigroup [with identity] and let $a \in S$. Then L(a), R(a), and J(a) are compact [connected].

1.27 Theorem. If S is a connected monoid, then each ideal of S is connected.

Proof. Let I be an ideal of S and let $a, b \in I$. Then J(a) and J(b) are connected, $J(a) \cap J(b) \neq \Box$, and $J(a) \cup J(b) \subset I$. We obtain that $J(a) \cup J(b)$ is a connected subset of I containing a and b, so I is connected. ■

1.28 Theorem. If S is a compact [connected] semigroup, then each minimal left ideal, each minimal right ideal, and each minimal ideal is compact [connected.]

Proof. Let L be a minimal left ideal of S and $x \in L$. Then $L = L(x) = Sx \cup x = Sx$, since L is minimal and Sx is a left ideal of S. The conclusion in both the compact and connected cases for minimal left ideals follow from $L = Sx$. Similar arguments work for minimal right ideals and minimal ideals. ■

Let S be any non-degenerate compact connected Hausdorff space with zero multiplication. Then any subset of S containing 0 is a [left, right] ideal. Hence, ideals of arbitrary compact connected semigroups need not be connected.

Observe at this point that none of our results guarantee the existence of minimal [left, right] ideals. Indeed, there are topological semigroups which have none, e.g., the semigroup $\mathbb{H}$. However, in the case of compact semigroups we have the following:

1.29 Theorem. Each compact semigroup contains a minimal [left, right] ideal.

Proof. Let S be a compact semigroup. It follows that the family of closed ideals of S is a descending family. Hence, $\cap \{I : I$ is a closed ideal of $S\} = J$ is an ideal of S. If K is any ideal of S, $K \subset J$, and $x \in K$, then SxS is a closed ideal of S which is contained in K. It follows that $J = M(S) \neq \square$.

To see that S has minimal left ideals, let Λ be the collection of all closed left ideals of S, partially ordered by inclusion. Again, each tower $\mathcal{T}$ in Λ has $\cap \mathcal{T}$ as lower bound in Λ and so, by Zorn's Lemma, minimal closed left ideals exist. Fix a minimal closed left ideal L and suppose that L' is a left ideal contained in L. If $x \in L'$, then Sx is a closed left ideal contained in L' and it follows that L is a minimal left ideal. A similar argument ensures the existence of minimal right ideals. ■

Clearly, the proof of the existence of minimal left ideals would work for minimal ideals. The separate proof for minimal ideals is given in 1.29 in order to emphasize that the axiom of choice is not required in that case.

1.30 Corollary. Let S be a compact semigroup. Then there exists an idempotent e in M(S) and for each such e, we have the following:

(a) Se is a minimal left ideal;

(b) eSe is a group;

(c) eS is a minimal right ideal; and

(d) SeS is the minimal ideal of S.

Proof. Since S is compact, these are equivalent by 1.23. By 1.29, the minimal ideal exists and, by 1.28, M(S) is a compact semigroup. Hence, there is an idempotent $e \in M(S)$. Clearly, SeS = M(S). ■

We now turn our attention to the other end of the spectrum of ideals and consider maximal proper ideals.

If S is a semigroup and $A \subset S$, then $[L_0(A), R_0(A)]$ $J_0(A)$ is the union of all [left, right] ideals of S contained in A. Observe that if A contains a [left, right] ideal of S, then $[L_0(A), R_0(A)]$ J_0 is a [left, right] ideal of S. Otherwise $[L_0(A) = \square, R_0(A) = \square]$ $J_0(A) = \square$.

The next result appears in [Koch and Wallace, 1954].

1.31 Theorem. Let S be a topological semigroup and let $A \subset S$. Then

(a) If A is closed, then $[L_0(A), R_0(A)]$ $J_0(A)$ is closed; and

(b) If S is compact and A is open, then $[L_0(A), R_0(A)]$ $J_0(A)$ is open.

Proof. The proofs are presented for J only. For (a), we assume $J_0(A) \neq \square$ and apply 1.24 to obtain that $\overline{J_0(A)}$ is an ideal. Hence, if A is closed, $\overline{J_0(A)} \subset A$ and it follows that $\overline{J_0(A)} = J_0(A)$.

For (b), if $J_0(A) \neq \square$, fix $x \in J_0(A)$. Then $J(x) = x \cup xS \cup Sx \cup SxS \subset A$, and by repeated application of 1.1, there exists an open set W containing x such that $J(W) \subset A$. Thus $x \in W \subset J(W) \subset J_0(A)$, and $J_0(A)$ is open. ■

1.32 Corollary. Let S be a compact semigroup, I a proper ideal of S, and $x \in S\backslash I$. Then $J_0(S\backslash x)$ is an open proper ideal of S.

An ideal I of a semigroup S is a *maximal proper ideal* of S if I is proper and is contained in no other proper ideal of S.

1.33 Theorem. Let S be a compact semigroup. Then each proper [left, right] ideal of S is contained in a maximal proper [left, right] ideal and each such is open.

Proof. Again, the proof is presented for ideals only. It follows immediately from 1.32 that each proper ideal of S is contained in an open ideal of S. Hence, each maximal proper ideal of S is open and any ideal of S which is maximal in the family $\mathcal{F}$ of each proper ideals of S containing an ideal J is actually maximal in the family of all proper ideals of S. Let $\mathcal{C}$ be a chain in $\mathcal{F}$ (partially ordered by inclusion) and let $C = \cup\, \mathcal{C}$. Then C is an open ideal of S containing J. Moreover, C is proper, for otherwise $\mathcal{C}$ would be towered open cover of S and the compactness of S would imply that S itself was a proper ideal. Hence $\mathcal{F}$ is inductive, and by Zorn's Lemma, $\mathcal{F}$ contains a maximal member. ∎

1.34 Theorem. Let S be a compact semigroup with identity 1 such that $S \neq H(1)$. Then $S\backslash H(1)$ is the unique maximal proper ideal of S. Moreover, if S is connected, then $S\backslash H(1)$ is dense in S.

Proof. It follows from 1.17 that $S\backslash H(1)$ is an ideal of S. That $S\backslash H(1)$ is the unique maximal proper ideal is then a consequence of 1.19 (e). In the case that S is connected we have that $\overline{S\backslash H(1)}$ is an ideal of S and, as $S\backslash H(1)$ is open, we have $\overline{S\backslash H(1)} \cap H(1) \neq \square$. Thus $\overline{S\backslash H(1)} = S$ and $S\backslash H(1)$ is dense in S. ∎

Recall that a continuum (compact connected Hausdorff space) is said to be *indecomposable* if it is not the union of two proper nondegenerate subcontinua.

The next result appears in [Wallace, 1955].

1.35 Theorem. Let S be a compact connected monoid. If S is indecomposable, then S is a group.

Proof. Let 1 denote the identity of S and assume that $S \neq H(1)$. Then $S \backslash H(1)$ is an ideal and so $M(S) \subset S \backslash H(1)$. Let V be an open set such that $M(S) \subset V \subset \overline{V} \subset S \backslash H(1)$. Then $J_0(V)$ is open and connected. Moreover, we have $M(S) \subset J_0(V) \subset \overline{J_0(V)} \subset S \backslash H(1)$. Now $S = \overline{J_0(V)} \cup \overline{S \backslash J_0(V)}$, and since S is indecomposable, we have that $S \backslash \overline{J_0(V)}$ is not connected. Let $S \backslash \overline{J_0(V)} = P \cup Q$ be a separation. Then, $\overline{J_0(V)} \cup P$ and $\overline{J_0(V)} \cup Q$ are proper non-degenerate subcontinua whose union is S, contrary to the indecomposability of S. ■

The following lemma will be of use in proving the corollary to our next theorem, as well as in other places in the text.

1.36 Lemma. Let S be a compact semigroup and let $\mathcal{F}$ and $\mathcal{G}$ be descending families of non-empty closed subsets of S. Then

(a) $[\cap \mathcal{F}][\cap \mathcal{G}] = \cap \{FG : F \in \mathcal{F} \text{ and } G \in \mathcal{G}\}$; and
(b) $[\cap \mathcal{F}]^2 = \cap \{F^2 : F \in \mathcal{F}\}$.

Proof. The containments from left to right are clear in both (a) and (b).

To complete the proof of (a), suppose that $x \in S \backslash [\cap \mathcal{F}][\cap \mathcal{G}]$. Then, using the compactness of $[\cap \mathcal{F}][\cap \mathcal{G}]$ and 1.1, we obtain open sets U and V containing $\cap \mathcal{F}$ and $\cap \mathcal{G}$ respectively, such that $x \in S \backslash UV$. Now, there exist $F \in \mathcal{F}$ and $G \in \mathcal{G}$ such that $F \subset U$ and $G \subset V$. Therefore, $x \in S \backslash FG \subset S \backslash \cap\{FG : F \in \mathcal{F} \text{ and } G \in \mathcal{G}\}$ and (a) is established.

To complete the proof of (b), it suffices to show that $\cap \{F^2 : F \in \mathcal{F}\} \subset \cap \{F_1F_2 : F_1 \in \mathcal{F} \text{ and } F_2 \in \mathcal{F}\}$. This, however, is an immediate consequence of the fact that $\mathcal{F}$ is a descending family. ■

1.37 Theorem. Let S be a compact semigroup. Then $S^2 = S$ if and only if $E(S) \cap (S \backslash I) \neq \square$ for each proper ideal I of S.

Proof. Suppose that $E(S) \cap (S \backslash I) \neq \square$ for each proper ideal I of S. Now S^2 is an ideal of S containing E, and hence $S^2 = S$.

Suppose, on the other hand, that $S^2 = S$ and let I be a proper ideal of S. By 1.33, there exists a maximal proper ideal J of S containing I. Letting $A = S\backslash J$, we obtain $SAS \cap A \neq \Box$, for otherwise $S = S^2 = S(S^2) = S^3 = SAS \cup SJS \subset J$. Fix $a \in A$. The maximality of J ensures that $S = J \cup SAS$ and so $a \in SAS$. If $B = \{x \in S : a \in xaS\}$, then B is closed. Moreover, B is a subsemigroup of S, since $x, y \in B$ implies $a \in xaS \subset x(yaS)S = xyaS$. Hence, B contains an idempotent e. Finally, $a \in eaS$ J implies $e \in S\backslash J$ implies $e \in S\backslash J \subset S\backslash I$. ∎

1.38 Corollary. If S is a compact semigroup, then $SES = \cap \{S^n\}_{n\in\mathbb{N}}$.

Proof. Observe that $(SES)^2 = SES$ and hence, premultiplying this equation by SES, we obtain $(SES)^3 = (SES)^2 = SES$. By induction, $(SES)^{n+1} = (SES)^n = SES$ for all $n \in \mathbb{N}$ and so $SES \subset S^n$. Let $T = \cap \{S^n\}_{n\in\mathbb{N}}$. Since $\cdots S^3 \subset S^2 \subset S$, $\{S^n : n \in \mathbb{N}\}$ is a tower of non-empty closed subsets of S. Hence, $T^2 = \cap \{S^{2n}\}_{n\in\mathbb{N}} = \cap \{S^n\}_{n\in\mathbb{N}} = T$. Now SES is an ideal of the compact semigroup T and $E \subset SES$. In view of 1.35, we see that $SES = T$. ∎

1.39 Corollary. Let S be a compact semigroup. Then $S^2 = S$ if and only if $S = SES$.

Proof. If $S^2 = 2$, then $S^n = S$ for all $n \in \mathbb{N}$ and so $S = \cap \{S^n\}_{n\in\mathbb{N}} = SES$. Conversely, if $S = SES$, then $S = SES = \cap \{S^n\}_{n\in\mathbb{N}} \subset S^2$ so that $S = S^2$. ∎

The next theorem is a characterization of compact connected semigroups S for which $S^2 = S$ and follows 1.39 rather naturally. It appears in [Koch and Wallace, 1958].

1.40 Theorem. Let S be a compact connected semigroup. Then $S^2 = S$ if and only if each dense [left, right] ideal [containing M(S)] is connected.

Proof. We present the proof for left ideals only. Suppose $S^2 = S$ and L is a dense left ideal containing M(S). Then $SL = S(L \cup M(S)) = \cup \{Sa\}_{a\in L} \cup M(S)$ and each set Sa meets the connected set M(S). It follows that SL is connected. Now, $SL \subset L \subset S$ and $\overline{SL} = S\overline{L} = S^2 = S$ so that L is connected.

Conversely, suppose that each dense left ideal containing M(S) is connected and $S^2 \neq S$. Let $a \in S\backslash S^2$ and let V be an open set containing a with $\overline{V} \cap S^2 = \Box$. Then $S\backslash F(V)$ (where F(V) denotes the boundary $\overline{V}\backslash V$ of V) is a dense ideal and is not connected. This contradiction yields the conclusion. ■

We now consider the concept of a Δ-ideal, which will have useful applications in later results.

If S is a semigroup and $\mathcal{U}$ is a non-empty subset of $S \times S$, then $\mathcal{U}$ is said to be a Δ-ideal provided $\Delta\mathcal{U} \cup \mathcal{U}\Delta \subset \mathcal{U}$, where the multiplication in $S \times S$ is coordinate-wise multiplication $((a,b),(c,d)) \to (ac,bd)$.

1.41 Theorem. Let S be a semigroup. Then:

(a) Δ is a Δ-ideal of S;

(b) If $\{A_i : i \in I\}$ is a family of Δ-ideals of S, then $\cup \{A_i\}_{i \in I}$ is a Δ-ideal of S;

(c) If $\mathcal{U}$ is a Δ-ideal of S, then so is $\mathcal{U} \cup \Delta$;

(d) If $\mathcal{U}$ is a Δ-ideal of S, then so is $\mathcal{U}^{-1}$;

(e) If $\mathcal{U}$ and $\mathcal{V}$ are Δ-ideals of S, then so is $\mathcal{U} \circ \mathcal{V}$; and

(f) If $\mathcal{U}$ is a Δ-ideal of S, then so is $Tr(\mathcal{U})$.

Proof. The proof of (a) is obvious.

For (b), we have $\Delta(\cup \{A_i\}_{i \in I}) = \cup \{\Delta A_i\}_{i \in I} \subset \{A_i\}_{i \in I}$ and similarly $(\cup \{A_i\}_{i \in I})\Delta \subset \cup \{A_i\}_{i \in I}$.

Now, (c) is a consequence of (a) and (b).

For (d), we have $\Delta\mathcal{U}^{-1} \cup \mathcal{U}^{-1}\Delta = (\mathcal{U}\Delta)^{-1} \cup (\mathcal{U}\Delta)^{-1} \subset \mathcal{U}^{-1}$.

To prove (e), fix $(x,x) \in \Delta$ and $(y,w) \in \mathcal{U} \circ \mathcal{V}$. Then there exists $z \in S$ such that $(y,z) \in \mathcal{U}$ and $(z,w) \in \mathcal{V}$. It follows that $(xy,xz) \in \mathcal{U}$ and $(xz,xw) \in \mathcal{V}$ and so $(xy,xw) \in \mathcal{U} \circ \mathcal{V}$. Therefore $\Delta(\mathcal{U} \circ \mathcal{V}) \subset \mathcal{U} \circ \mathcal{V}$ and similarly, $(\mathcal{U} \circ \mathcal{V})\Delta \subset \mathcal{U} \circ \mathcal{V}$.

Finally, (f) follows from (e) by induction. ■

1.42 Theorem. Let S be a semigroup and let $\mathcal{V}$ be a subset of $S \times S$ containing Δ. Then $\mathcal{U} = \cup\{\mathcal{W} : \mathcal{W}$ is a Δ-ideal and $\mathcal{W} \subset \mathcal{V}\}$ is a Δ-ideal. Moreover,

(a) If S is a topological semigroup and $\mathcal{V}$ is closed, then $\mathcal{U}$ is closed; and

(b) If S is a compact semigroup and $\mathcal{V}$ is open, then $\mathcal{U}$ is open.

Proof. Since $\Delta \subset \mathcal{V}$, we have $\mathcal{U} \neq \square$. Now, $\mathcal{U}$ is a Δ-ideal of S in view of 1.41 (b).

For (a), suppose S is a topological semigroup and $\mathcal{V}$ is closed. Then, as coordinate-wise multiplication is continuous, we obtain $\Delta\overline{\mathcal{U}} \cup \overline{\mathcal{U}}\Delta \subset \overline{\Delta\mathcal{U}} \cup \overline{\mathcal{U}\Delta} \subset \overline{\mathcal{U}}$. Thus $\overline{\mathcal{U}}$ is a Δ-ideal and, as $\mathcal{V}$ is closed, $\overline{\mathcal{U}} \subset \mathcal{V}$. It follows that $\overline{\mathcal{U}} = \mathcal{U}$.

To prove (b), suppose that S is a compact semigroup and $\mathcal{V}$ is open. Fix $(x,y) \in \mathcal{U}$. Then $(x,y) \cup (x,y)\Delta \cup \Delta(x,y) \cup \Delta(x,y)\Delta \subset \mathcal{V}$, and hence by repeated applications of 1.1, there is an open set V containing (x,y) such that $V \cup V\Delta \cup \Delta V \cup \Delta V\Delta \subset \mathcal{V}$. This set is obviously a Δ-ideal and hence $(x,y) \in V \subset \mathcal{U}$. ■

If S is a [topological] semigroup and X is a set [Hausdorff space], then S is said to *act* on X as a semigroup of transformations by the action α provided $\alpha : S \times X \to X$ is a [continuous] function such that $\alpha(st,x) = \alpha(s,\alpha(t,x))$ for all $s, t \in S$ and all $x \in X$. If $T \subset S$ and $\square \neq A \subset X$, then A is T-*invariant* (or a T-*ideal*) provided $\alpha(T \times A) \subset A$. Ordinarily, α is suppressed and we simply write $S \times X \to X$ is an act.

We now illustrate how one can consider the types of ideals we have discussed as invariant subsets of certain actions. Let S be a semigroup; denote by D the set S with multiplication $s \circ t = ts$, and let $A \subset S$ and $\mathcal{U} \subset S \times S$. Then:

(i) A is a left ideal if and only if A is an S^1-ideal of the action $\alpha : S^1 \times S \to S$ defined by $\alpha(s,x) = sx$;

(ii) A is a right ideal if and only if A is a D^1-ideal of the action $\beta : D^1 \times S \to S$ defined by $\beta(s,x) = xs$;

(iii) A is an ideal if and only if A is an $S^1 \times D^1$-ideal of the action $\gamma : (S^1 \times D^1) \times S \to S$ defined by $\gamma((s,t),x) = sxt$; and

(iv) $\mathcal{U}$ is a Δ-ideal if and only if $\mathcal{U}$ is an $S^1 \times D^1$-ideal of the action $\delta : (S^1 \times D^1) \times (S \times S) \to S \times S$ defined by $\delta((s,t),(x,y)) = (sxt,syt)$.

Finally, we observe that the relative ideals introduced in [Wallace, 1962] can also be considered as special cases of act ideals. If $T \subset S$, then a subset A of S is a *left* T-*ideal* if $A \neq \square$ and $TA \subset A$. Observe that S is a left T-ideal if and only if A is a T-ideal of the action α in (i). These observations are made in [Day, 1971] and the definition of an act ideal appears in [Day and Wallace, 1967a].

It is not surprising that many results about ideals and Δ-ideals extend to results about act ideals. We present only one such result which generalizes 1.31(b) and 1.42(b). For a comprehensive treatment of acts, the reader is referred to [Aczel and Wallace, 1967], [Balman, 1966], [Bednarek and Wallace, 1966; 1967a,b,c], [Borrego, 1969; 1970], [Borrego and DeVun, 1970; 1971; 1972], [Day, 1968; 1971], [Day and Hofmann, 1972], [Day and Wallace, 1967a,b], [Friedberg, 1972], [Hanson, 1971a,b; 1972a,b], [Hanson and Kind, 1974], [Keleman, 1969; 1972; 1973], [King, 1972], [Lin, 1970], [McGranery, 1974], [Norris, 1971], [Reilly, 1974], [Sheldon, 1970], [Stadtlander, 1968; 1969; 1970; 1971; 1974], [Taylor, 1971], and [Wallace, 1955; 1957; 1962; 1963; 1967; 1969]. Also, a sizable literature is available on finite acts. For this topic, the reader is referred to the work of K. Krohn, John Rhodes, and B. R. Tilson.

1.43 Theorem. Let S be a compact semigroup, T a closed subsemigroup of S, X a compact Hausdorff space, A a subset of X, and $\alpha : S \times X \to X$ an act. If A is a T-ideal and V is an open subset of X containing A, then there exists an open T-ideal U such that $A \subset U \subset V$.

Proof. Let U be the union of all T-ideals contained in V. Then clearly U is a T-ideal and $A \subset U \subset V$. We show that U is open. Let

$x \in U$. Then $(T \times \{x\}) \subset V$ and by 1.6, there exists an open set W containing x such that $\alpha(T \times W) \subset V$. Since V is open and $x \in V$, we can assume that $W \subset V$. Now, $W \cup \alpha(T \times W)$ is a T-ideal contained in V and hence $x \in W \subset U$. ■

We conclude this section by presenting some examples of ideals in topological semigroups.

Let J be an ideal of I_u (or I_n or I_m). Then J is connected. Let $a = \sup\{x \in I : x \in J\}$. Then if J is closed, we have that $J = [0,a]$ and $J = [0,a)$ otherwise. We obtain that each closed ideal of I_u (or I_n or I_m) is principal.

Let I be an ideal of $\mathbb{D}$. Then I is connected. Let $r = \sup\{|x| : x \in I\}$. Then if I is closed, we have that $I = \{x \in \mathbb{D} : |x| \leq r\}$ and $I = \{x \in \mathbb{D} : |x| < r\}$ otherwise. As in the preceding paragraph, each closed ideal of $\mathbb{D}$ is principal. Observe in the preceding example that each principal ideal has exactly one generator, but in this example if $I = J(x)$, then $I = J(y)$ for each $y \in \mathbb{D}$ such that $|x| = |y|$.

Let T be the Triangle. For $x \neq 0$, the principal left and right ideals are illustrated with the presentation of the Triangle example. Since T has an identity, $L((x,y)) = T(x,y) = J((x,y))$ and $R((x,y)) = (x,y)T$. Observe that $M(T) = \{(0,y) : 0 \leq y \leq 1\}$, and M(T) consists of left zero elements.

HOMOMORPHISMS AND CONGRUENCES

In this section we present some of the basic ideas involved in mappings between topological semigroups, and identifying one topological semigroup with another. We introduce the concepts of homomorphisms and congruences in their algebraic setting and analogs of these concepts in topological semigroup.

If S and T are semigroups, a function $\phi : S \to T$ is called a *homomorphism* if $\phi(xy) = \phi(x)\phi(y)$ for each $x, y \in S$. If ϕ is surjective (onto), then ϕ is called a *surmorphism*. If ϕ is also injective (one-to-one), then ϕ is called an *algebraic isomorphism* and S and T are said to be *algebraically isomorphic*.

If S and T are topological semigroups and $\phi : S \to T$ is both an algebraic isomorphism and a homeomorphism, then ϕ is called a *topological isomorphism* and S and T are said to be *topologically isomorphic*.

If $\phi : \mathbb{H}^* \to I_u$ were defined by

$$\phi(x) = \begin{cases} \exp(-x) & \text{if } x \neq \infty; \\ 0 & \text{if } x = \infty. \end{cases}$$

then ϕ would be a topological isomorphism of $\mathbb{H}^*$ onto I_u.

It is clear that a composition of [continuous] homomorphisms and a restriction of a [continuous] homomorphism to a subsemigroup is again a [continuous] homomorphism.

If S and T are [topological] semigroups, an algebraic [topological] isomorphism of S onto a subsemigroup of T is called an *algebraic* [*topological*] embedding of S into T.

Observe that if S is a subsemigroup of a [topological] semigroup T, then the inclusion map of S into T is an algebraic [topological] embedding of S into T.

It is simple to show that if $\phi : S \to T$ is a homomorphism, then ϕ preserves subsemigroups and subgroups, i.e., if A is a subsemigroup [subgroup] of S, then $\phi(A)$ is a subsemigroup [subgroup] of T. In the case that ϕ is a surmorphism, then ϕ preserves ideals and minimal ideals of all three types and ϕ^{-1} preserves subsemigroups, left ideals, right ideals, and ideals.

If $\phi : S \to T$ is a homomorphism of a semigroup S into a semigroup T, then ϕ preserves idempotent elements. For an idempotent e in T, it may be the case that $\phi^{-1}(e)$ contains no idempotents (even if ϕ is a homomorphism, e.g. if $S = \mathbb{N}$, $T = \{e\}$, and $\phi : \mathbb{N} \to \{e\}$ is the trivial map). However, in the case that S and T are compact and ϕ is continuous, we have the following:

1.44 Theorem. Let S and T be compact semigroups and let $\phi : S \to T$ be a continuous surmorphism. Then $\phi(E(S)) = E(T)$.

Proof. It is clear that $\phi(E(S)) \subset E(T)$. Let $e \in E(T)$. Then $\phi^{-1}(e)$ is a closed, and hence compact, subsemigroup of S. By 1.8, $\phi^{-1}(e)$ contains an idempotent u. It follows that $\phi(u) = e$ and $E(T) \subset \phi(E(S))$. ∎

We turn now to a discussion of congruences on semigroups, and the connection between these and homomorphisms.

A relation R on a semigroup S is said to be *left* [*right*] *compatible* if $(a,b) \in R$ and $x \in S$ implies that $(xa,xb) \in R$ [$(ax,bx) \in R$], and *compatible* if R is both left and right compatible. Note that a non-empty relation R on S is compatible if and only if it is a Δ-ideal.

A compatible equivalence on a semigroup S is called a *congruence* on S.

Observe that an equivalence R on a semigroup S is a congruence if and only if $(a,b) \in R$ and $(c,d) \in R$ imply $(ac,bd) \in R$. It is also a very useful observation (for constructing examples) that each equivalence on a [left, right] zero semigroup is a congruence.

Observe that the intersection of any collection of congruences on a semigroup S is again a congruence on S and that $S \times S$ is itself a congruence on S. For $A \subset S \times S$, the *congruence on* S *generated by* A is the intersection of all congruences on S containing A. If $Eq(A) = Tr(A \cup A^{-1} \cup \Delta)$, then Eq(A) is the *equivalence on* S *generated by* A, and $Cong(A) = Tr(Eq(A) \cup \Delta Eq(A) \cup Eq(A)\Delta \cup \Delta Eq(A)\Delta)$ is the congruence on S generated by A.

If R is an equivalence on a set X and $x \in X$, then $\{y \in X : (x,y) \in R\}$ is called the R-*class* of X containing x. Recall that the R-classes of X are disjoint and X is the union of its R-classes. The set X/R of R-classes is called the *quotient of* X *mod* R, and the function $\pi : X \to X/R$ which assigns to each x in X the R-class containing x is called the *natural map*. Observe that for each $x \in X$, the set $\pi^{-1}\pi(x)$ is the R-class of x.

If R is an equivalence on a set S, then for the remainder of this section we denote the natural map by $\pi : X \to X/R$. Observe that π is surjective.

1.45 Theorem. Let S be a semigroup and let R be a congruence on S. Then S/R is a semigroup under multiplication defined by $(\pi(x),\pi(y)) \mapsto \pi(xy)$, and $\pi : S \to S/R$ is a surmorphism.

Proof. To see that multiplication is well-defined on S/R, suppose that $a \in \pi^{-1}\pi(x)$ and $b \in \pi^{-1}\pi(y)$. Then $(a,x) \in R$ and $(b,y) \in R$, so that $(ab,xy) \in R$, and hence $ab \in \pi^{-1}\pi(xy)$, i.e., $\pi(ab) = \pi(xy)$. Associativity of multiplication on S/R follows from that of S. It is clear that π is a surmorphism. ∎

It is useful to observe that if R is a congruence on a semigroup S, and m and m' are the multiplications on S and S/R, respectively, then m' is the unique multiplication on S/R such that the following diagram commutes:

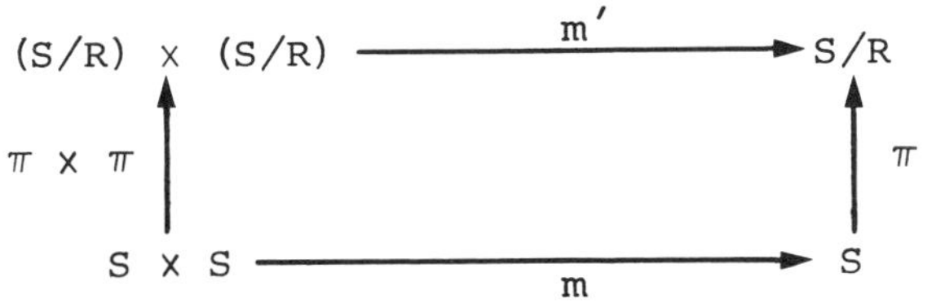

1.46 Theorem. Let S be a semigroup, I an ideal of S, and $R = (I \times I) \cup \Delta$. Then R is a congruence on S.

Proof. It is clear that R is an equivalence. Now $I \times I$ is a Δ-ideal and hence R is a Δ-ideal. Thus, R is a congruence on S. ∎

If S is a semigroup and I is an ideal of S, then the semigroup $S/((I \times I) \cup \Delta)$ is called the *Rees quotient semigroup of* S *mod* I and is denoted S/I.

Observe that the image of I under the natural map is a zero for the semigroup S/I. Intuitively, the underlying set on which the semigroup S/I is defined is the one obtained from S by "shrinking" I to a point. We will obtain a topological analog of the Rees quotient semigroup for locally compact σ-compact semigroups.

If R is an equivalence [congruence] on a topological space [semigroup] X, we will call R a *closed equivalence* [*congruence*] if R is a closed subset of $X \times X$.

We have observed that it was straightforward to obtain a description of the smallest transitive relation containing a subset A of $S \times S$ (Tr(A)), the smallest equivalence containing A(eq(A)), and the smallest congruence containing A(Cong(A)). The problem of finding descriptions of the smallest closed relations of these three types in case S is a topological semigroup is much more difficult. The reason for this is that the closure of a transitive relation need not be transitive. In fact, the closure of a congruence need not be transitive. For example, let S be the following subspace of $[0,2] \times [0,1]$: $S = \{(0,\frac{1}{n}) : n \in \mathbb{N}\} \cup \{(1 - \frac{1}{n},\frac{1}{n}) : n \in \mathbb{N}\} \cup \{(1 + \frac{1}{n},\frac{1}{n}) : n \in \mathbb{N}\} \cup \{(2,\frac{1}{n}) : n \in \mathbb{N}\} \cup \{(0,0),(1,0),(2,0)\}$ with left zero multiplication. Let A be the equivalence on S whose classes are $\{(0,\frac{1}{n}),(1 - \frac{1}{n},\frac{1}{n})\}$ for $n \in \mathbb{N}$, $\{(1 + \frac{1}{n},\frac{1}{n}),(2,\frac{1}{n})\}$ for $n \in \mathbb{N}$, $\{(0,0)\}$, $\{(1,0)\}$, and $\{(2,0)\}$. The compact semigroup S and the classes of A are illustrated in the following figure:

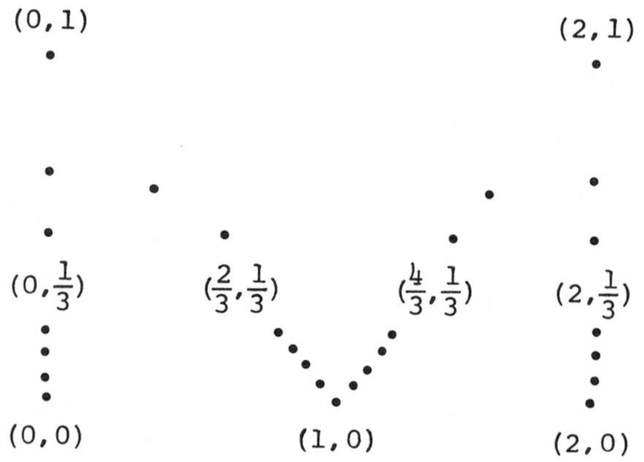

Observe that A is a congruence on S, $((0,0),(1,0)) \in \overline{A}$, $((1,0),(2,0)) \in \overline{A}$, but $((0,0),(2,0)) \notin \overline{A}$, so that $\overline{A}$ is not transitive.

It is interesting to note that, if S is a compact semigroup and A is a Δ-ideal, then the smallest closed equivalence containing A is a congruence. To see this, one lets C be the smallest closed equivalence containing A and $\mathcal{A} = \{B : A \subset B \subset C$ and B is a Δ-ideal$\}$. Then one shows $\cup \mathcal{A} = C$ is a congruence [Carruth, Hofmann, and Mislove, 1972]. There are transfinite methods for obtaining the smallest closed congruence containing a subset of $S \times S$ [Hofmann and Mostert, 1966]. However, they are quite involved and, having pointed out the difficulties that one confronts, we do not go into the details.

If $f : X \to Y$ is a function from a space X into a space Y and $A \subset X$, then A is said to be *f-saturated* if $f^{-1}f(A) = A$. If $f : X \to Y$ is surjective, then f is said to be a *quotient map* if W being open [closed] in Y is equivalent to $f^{-1}(W)$ being open [closed] in X. Observe that a quotient map is necessarily a continuous surjection. It is clear that each open continuous surjection and each closed continuous surjection of X onto Y is a quotient map.

If X is a space, Y is a set, and $f : X \to Y$ is a surjection from X onto Y, then we can define a topology on Y by declaring a subset W of Y to be open if and only if $f^{-1}(W)$ is open in X. This topology is called the *quotient topology on* Y *induced by* f. Note that if Y is given the quotient topology induced by f, then f is a quotient map.

If R is an equivalence on a space X, we give X/R the quotient topology induced by the natural map $\pi : X \to X/R$.

The following lemma will be used in establishing the topological versions of our next three theorems and later results in this section.

1.47 Lemma. Let X, Y, and Z be spaces and let f, g and h be functions such that the following diagram commutes:

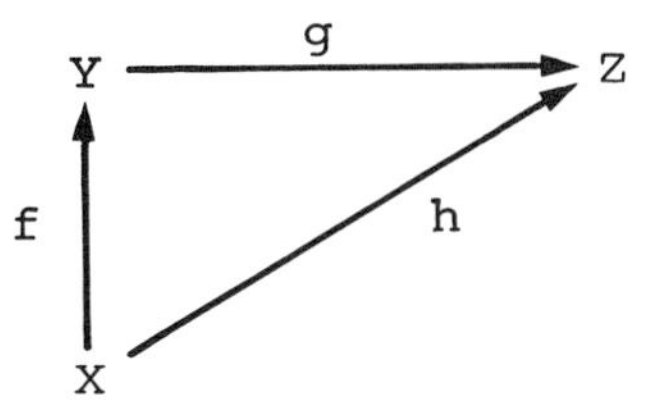

Then:

(a) If f is quotient and h is continuous, then g is continuous; and

(b) If both f and h are quotient, then g is quotient.

Proof. To prove (a), let U be an open subset of Z. Then, since h is continuous, we have $h^{-1}(U) = f^{-1}(g^{-1}(U))$ is open in X. Since f is quotient, $g^{-1}(U)$ is open in Y, and hence g is continuous.

To prove (b), first observe that the continuity of g follows from (a). Let U be a subset of Z such that $g^{-1}(U)$ is open in Y. Then $h^{-1}(U) = f^{-1}(g^{-1}(U))$ is open since f is continuous. That U is open now follows from the hypothesis that h is quotient. ■

If S and T are semigroups and $\phi : S \to T$ is a homomorphism, we denote by $K(\phi)$ the relation $\{(x,y) \in S \times S : \phi(x) = \phi(y)\}$.

1.48 Induced Homomorphism Theorem. Let A, B, and C be [topological] semigroups, $\alpha : A \to B$ a [quotient] surmorphism, and $\beta : A \to C$ a [continuous] homomorphism such that $K(\alpha) \subset K(\beta)$. Then there exists a unique [continuous] homomorphism $\gamma : B \to C$ such that the diagram:

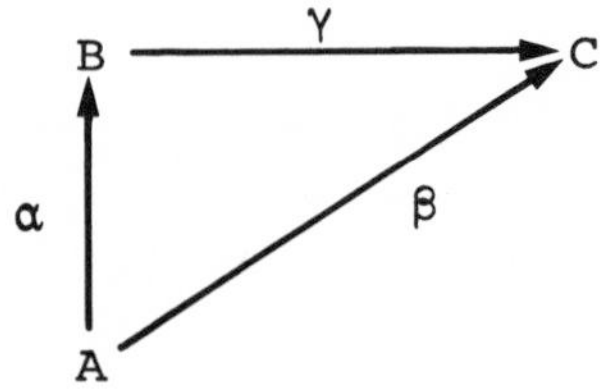

commutes.

Proof. Define $\gamma(x) = \beta(\alpha^{-1}(x))$ for each $x \in B$. Then $\gamma : B \to C$ is well defined, since $K(\alpha) \subset K(\beta)$. It is straightforward to show that γ is a homomorphism. In the topological case, since β is continuous and α is quotient, γ is continuous by 1.47 (a). The uniqueness of γ is clear. ■

1.49 First Isomorphism Theorem. Let S and T be semigroups and let $\phi : S \to T$ be a surmorphism. Then $K(\phi)$ is a congruence on S and there exists a unique algebraic isomorphism $\psi : S/K(\phi) \to T$ such that the diagram:

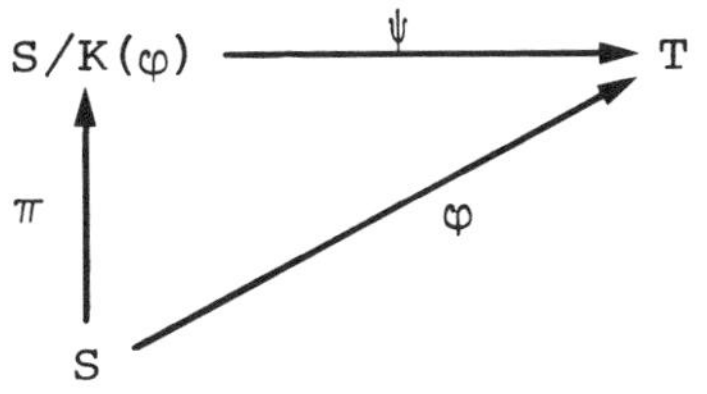

commutes. Moreover, if S and T are topological semigroups and $\phi : S \to T$ is a continuous surmorphism, then $K(\phi)$ is a closed congruence on S and these are equivalent:

(a) ψ^{-1} is continuous;

(b) ψ is a topological isomorphism; and

(c) ϕ is quotient.

Finally, if these equivalent statements hold, then $S/K(\phi)$ is a topological semigroup.

Proof. It is a straightforward argument that $K(\phi)$ is a congruence on S. Using [1.47 (a)], 1.49, and the fact that $K(\pi) = K(\phi)$, we obtain a [continuous] homomorphism such that the diagram commutes. Again, it is straightforward to show that ψ is an algebraic isomorphism and that ψ is unique relative to the commutativity of the diagram.

In the topological case, $K(\phi)$ is closed, since $K(\phi) = (\phi \times \phi)^{-1}(\Delta(T))$.

That (a) implies (b) is immediate.

The fact that (b) implies (c) is immediate from 1.47 (b), since both ψ and π are quotient maps.

That (c) implies (a) follows from 1.47 (a) and the fact that ϕ is quotient.

The last assertion of the theorem is transparent. ∎

If $S/K(\phi)$ in 1.49 is a topological semigroup, this does not insure that ψ is a topological isomorphism. The following example illustrates this fact:

Let S denote the discrete multiplicative semigroup of non-negative integers $\{0,1,2,\ldots\}$ and T the compact subsemigroup $\{\frac{1}{n} : n \in \mathbb{N}\} \cup \{0\}$ of I_u (the usual interval). Define $\phi : S \to T$ by

$$\phi(n) = \begin{cases} 0 & \text{if } n = 0; \\ 1/n & \text{if } n \neq 0. \end{cases}$$

Then ϕ is a continuous surmorphism, $K(\phi) = \Delta(S)$, and hence $S/K(\phi)$ is a topological semigroup which is topologically isomorphic to S (and hence not to T). Since $\phi^{-1}(0) = \{0\}$ is open in S, and $\{0\}$ is not open in T, we have that ϕ is not quotient. In view of 1.49, we see that $\psi : S/K(\phi) \to T$ is not a topological isomorphism.

1.50 Second Isomorphism Theorem. Let S be a [topological] semigroup and let R_1 and R_2 be [closed] congruences on S such that [S/R_1 and S/R_2 are topological semigroups and] $R_1 \subset R_2$. Then there exists a [closed] congruence T on S/R_1 [such that $(S/R_1)/R$ is a topological semigroup] and an algebraic [topological] isomorphism $\psi : (S/R_1)/R \to S/R_2$ such that the diagram:

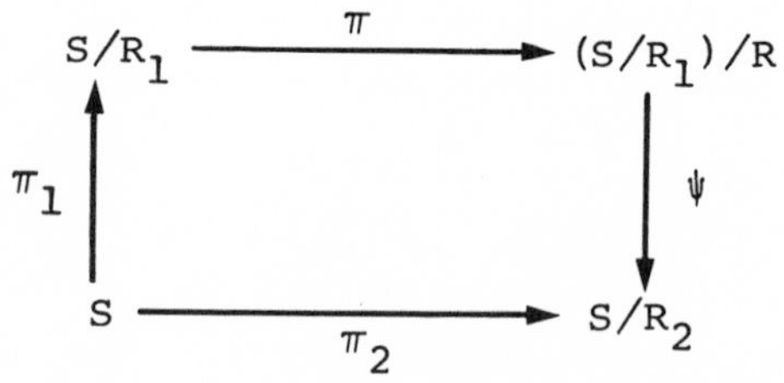

commutes, where π_1, π_2, and π are natural maps.

Proof. In view of 1.48 and the fact that $R_1 \subset R_2$, there exists a unique [continuous] homomorphism ϕ such that the diagram:

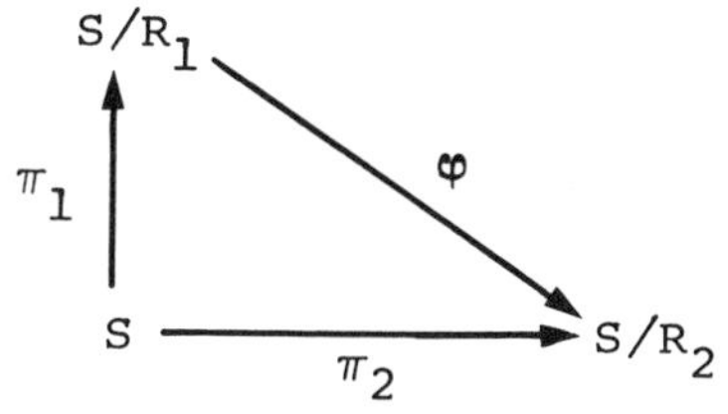

commutes. Now, 1.49 implies the existence of an algebraic isomorphism ψ such that the following diagram commutes:

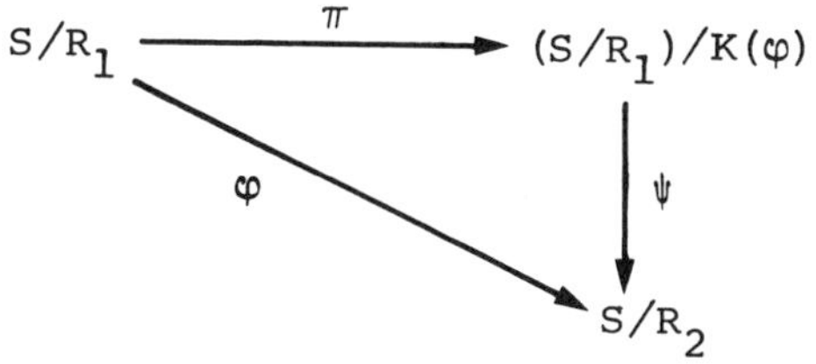

We simply let $R = K(\phi)$ and observe that ψ is a topological isomorphism in the topological case since ϕ is quotient (1.47 (b)). This completes the proof, however, we remark that $R = (\pi_1 \times \pi_1)(R_2)$. ∎

There are two major factors in 1.48, 1.49, and 1.50 which separate the comparatively simple algebraic results from their topological analogs. The first is that the surmorphisms in 1.48 and 1.49 which appear in the hypothesis are quotient in the topological version. To obtain a general topological analog of the algebraic results in these theorems, it would be desirable to replace the condition "quotient" by "continuous surmorphism". However, as illustrated above, this replacement cannot generally be made, even for locally compact σ-compact semigroups. In the case of compact semigroups it is simple to see that this replacement can be made, since continuous surmorphisms of compact semigroups are closed and hence quotient. The second factor is the requirement that S/R be a topological semigroup when S is a topological semigroup and R is a closed congruence on S. This condition appears in 1.50 and jointly with the quotient hypothesis in 1.49. We will prove that it is sufficient for S to be locally compact and σ-compact.

Observe that if R is a congruence on a topological semigroup S such that S/R is a topological semigroup, then the natural map $\pi : S \to S/R$ is a continuous homomorphism and $R = K(\pi)$. The requirement that S/R be Hausdorff forces $R = (\pi \times \pi)^{-1}(\Delta(S/R))$ to be closed.

We now present an example of a topological semigroup S and a closed congruence R on S such that S/R is not a topological semigroup, since S/R fails to be Hausdorff. We will see an example later where S/R is Hausdorff but multiplication on S/R fails to be continuous.

Let S be a Hausdorff space which is not normal and define multiplication on S by $(x,y) \mapsto x$. Then S is a topological semigroup. Let A and B be two disjoint closed subsets of S such that each open set containing A meets each open set containing B, and let $R = \Delta(S) \cup (A \times A) \cup (B \times B)$. Then R is closed congruence on S but S/R is not a topological semigroup, since S/R is not Hausdorff. Observe that multiplication on S/R is continuous, since it is left zero multiplication.

We now proceed to establish a sequence of lemmas which will be utilized to demonstrate that if S is a locally compact σ-compact semigroup and R is a closed congruence on S, then S/R is a topological semigroup. This result was established for compact semigroups in [Wallace, 1955] and extended to locally compact σ-compact semigroups in [Lawson and Madison, 1971].

1.51 Lemma. Let S be a topological semigroup and let R be a closed congruence on S such that $\pi \times \pi : S \times S \to (S/R) \times (S/R)$ is a quotient map. Then S/R is a topological semigroup.

Proof. Let m denote multiplication on S and m' multiplication on S/R. Then the diagram:

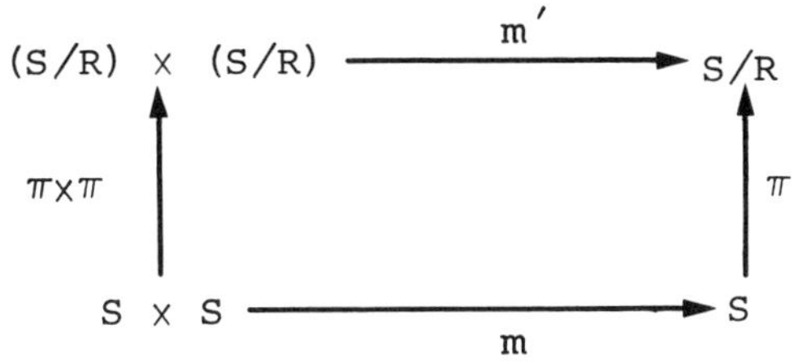

commutes. Since $\pi \circ m$ is continuous and $\pi \times \pi$ is quotient, m' is continuous.

Since $R = (\pi \times \pi)^{-1}(\Delta(S/R))$ is closed and $\pi \times \pi$ is quotient, $\Delta(S/R)$ is closed. Hence S/R is Hausdorff. ■

1.52 Lemma. Let R be a closed equivalence on a space X and let K be a compact subset of X. Then $\pi^{-1}\pi(K)$ is closed.

Proof. Let $\pi_1 : X \times K \to X$ denote first projection. Then $\pi^{-1}\pi(K) = \pi_1((X \times K) \cap R)$ is closed. ■

1.53 Lemma. Let X be a compact Hausdorff space and let R be a closed equivalence on S. Then:

(a) $\pi : X \to X/R$ is closed;

(b) If V is an open subset of X, then $V_R = \{x \in X : \pi^{-1}\pi(x) \subset V\}$ is open and π-saturated;

(c) If A is a closed π-saturated subset of X and W is an open subset of X such that $A \subset W$, then there exists an open π-saturated subset U and a closed π-saturated subset K such that $A \subset U \subset K \subset W$; and

(d) X/R is Hausdorff.

Proof. To prove (a), let C be a closed subset of X. Then C is compact, and hence $\pi^{-1}\pi(C)$ is closed. Since π is quotient, $\pi(C)$ is closed, so π is closed.

To prove (b), observe that $V_R = V \backslash \pi^{-1}\pi(X \backslash V)$. Since π is closed, we have that V_R is open. It is clear that V_R is π-saturated.

To prove (c), observe that $A \subset W_R \subset W$, since A is π-saturated. Since X is normal and W_R is open, there exists an open set V such that $A \subset V \subset \overline{V} \subset W_R \subset W$. Let $U = V_R$ and $K = \pi^{-1}\pi(V)$. Then U and K satisfy the required conditions.

To prove (d), let $\pi(x)$ and $\pi(y)$ be distinct points of X/R. Then $\pi^{-1}\pi(x)$ and $\pi^{-1}(y)$ are disjoint closed subsets of X. Since X is normal, there exist disjoint open sets M and N containing $\pi^{-1}\pi(x)$ and $\pi^{-1}\pi(y)$, respectively. We obtain that $\pi(M_R)$ and $\pi(N_R)$ are disjoint open subsets of X/R containing $\pi(x)$ and $\pi(y)$, respectively. ■

Although the next theorem is subsumed in 1.56, we present its proof here, since it is so accessible with the lemmas established thus far. This result is found in [Wallace, 1955].

1.54 Theorem. Let S be a compact semigroup and let R be a closed congruence on S. Then S/R is a compact semigroup.

Proof. By 1.53 (d), S/R is Hausdorff. Since $S \times S$ is compact, $(S/R) \times (S/R)$ is Hausdorff, and $\pi \times \pi$ is continuous, we have that $\pi \times \pi$ is closed and hence quotient. The conclusion follows from 1.51. ■

1.55 Lemma. Let X be a locally compact σ-compact space. Then there exists a sequence $\{K_n\}$ of compact subsets of X such that $X = \cup \{K_n\}_{n \in \mathbb{N}}$ and $K_n \subset K^0_{n+1}$ (the interior of K_{n+1} in X) for each $n \in \mathbb{N}$.

Proof. Let $X = \cup \{C_n\}_{n\in\mathbb{N}}$, where C_n is compact for each $n \in \mathbb{N}$. Let $K_1 = C_1$ and $\mathcal{K}_1$ a finite cover of K_1 by open sets with compact closures. Let $K_2 = \overline{\cup \mathcal{K}_1}$. Since $K_2 \cup C_2$ is compact, there exists a finite cover $\mathcal{K}_2$ of $K_2 \cup C_2$ by open sets with compact closures. Let $K_3 = \overline{\cup \mathcal{K}_2}$. Continuing recursively, let K_{n+1} be a compact subset of X such that $K_n \cup C_n \subset K^0_{n+1}$. ■

The following result appears in [Lawson and Madison, 1971].

1.56 Theorem. Let S be a locally compact σ-compact semigroup and let R be a closed congruence on S. Then S/R is a topological semigroup.

Proof. We will establish that S/R is a topological semigroup by proving that $\pi \times \pi : S \times S \to (S/R) \times (S/R)$ is quotient and applying 1.51.

Let Q be a subset of $(S/R) \times (S/R)$ such that $(\pi \times \pi)^{-1}(Q)$ is open in $S \times S$. We want to show that Q is open in $(S/R) \times (S/R)$. Let $(\pi(a),\pi(b)) \in Q$. Then $(\pi^{-1}\pi(a)) \times (\pi^{-1}(b)) \subset (\pi \times \pi)^{-1}(Q)$. By 1.55, there exists a sequence $\{K_n\}$ of compact subsets of S such that $S = \cup \{K_n\}_{n\in\mathbb{N}}$ and $K_n \subset K^0_{n+1}$ for each $n \in \mathbb{N}$. Assume, without loss of generality, that both a and b lie in K_1. Let $V_0 = \pi^{-1}\pi(a) \cap K_1$ and $W_0 = \pi^{-1}\pi(b) \cap K_1$. By 1.1, there exist sets G and T which are open in K_1 such that $V_0 \times W_0 \subset G \times T \subset (\pi \times \pi)^{-1}(Q) \cap (K_1 \times K_1)$. Using 1.53 (c) and the fact that K_1 is compact, we obtain sets V_1 and W_1 which are closed in K_1 (and hence compact) and $\pi|K_1$-saturated such that V_1 is a K_1-neighborhood of V_0, W_1 is a K_1-neighborhood of W_0, $V_1 \subset G$, and $W_1 \subset T$. By 1.52, $\pi^{-1}\pi(V_1)$ and $\pi^{-1}\pi(W_1)$ are closed. Again by 1.1, there exists sets M and N which are open in K_2 so that $(\pi^{-1}\pi(V_1) \cap K_2) \times (\pi^{-1}\pi(W_1) \cap K_2) \subset M \times N \subset (\pi \times \pi)^{-1}(Q) \cap (K_2 \times K_2)$. As above there are sets V_2 and W_2 which are closed in K_2 (and hence compact) and $\pi|K_2$-saturated such that V_2 is a K_2-neighborhood of $\pi^{-1}\pi(V_1) \cap K_2$, W_2 is a K_2-neighborhood of $\pi^{-1}\pi(W_1) \cap K_2$, $V_2 \subset M$, and $W_2 \subset N$. Define recursively a pair of towered sequences $\{V_n\}$ and $\{W_n\}$ such that V_n and W_n are closed in K_n (and hence compact) and $\pi|K_n$-saturated such that V_n is a K_n-neighborhood of $\pi^{-1}\pi(V_{n-1}) \cap K_n$, W_n is a K_n-neighborhood of $\pi^{-1}\pi(W_{n-1}) \cap K_n$, and $V_n \times W_n \subset (\pi \times \pi)^{-1}(Q)$ for each $n \in \mathbb{N}$.

Let $V = \cup \{V_n\}_{n\in\mathbb{N}}$ and $W = \cup\{W_n\}_{n\ \mathbb{N}}$. It is clear that $\pi^{-1}\pi(a) \subset V$ and $\pi^{-1}\pi(b) \subset W$. Observe that V is π-saturated, since $V = \cup \{\pi^{-1}\pi(V_n)\}_{n\in\mathbb{N}}$, and similarly W is π-saturated.

To see that V is open, let $x \in V$. Then $x \in V_n$ for some $n \in \mathbb{N}$. Now V_{n+1} is a K_{n+1}-neighborhood of V_n and $V_n \subset K^0_{n+1}$. It follows that $V_{n+1} \cap K^0_{n+1}$ is a K^0_{n+1}-neighborhood of V_n. Since K^0_{n+1} is open in S, $V_{n+1} \cap K^0_{n+1}$ is a neighborhood of V_n. Observing that $V_n \subset V_{n+1} \cap K^0_{n+1} \subset V_{n+1} \subset V$, we see that V is a neighborhood of x, and hence V is open. A similar argument proves that W is open. Finally, $(\pi(a),\pi(b)) \in \pi(V) \times \pi(W) \subset Q$ and $\pi(V) \times \pi(W)$ is open, since V and W are open π-saturated subsets of S. ■

The conclusion of 1.56 holds if "locally compact σ-compact" is replaced by "k_ω" [Lawson and Madison, 1971]. A k_ω-space is a space which is the quotient image of a locally compact σ-compact space. Lawson and Madison proved that each topological semigroup on a k_ω-space is the quotient image of a locally compact σ-compact semigroup. Although it is not our intention to prove 1.56 for semigroups on k_ω-spaces, one could obtain the result by combining this theorem with 1.50.

In the conclusion of 1.56, observe that no claim is made that S/R inherits the topological properties of S (except Hausdorff). However, in view of the preceding paragraph, we see that S/R is k_ω. The following example illustrates the fact that S/R need not be locally compact.

Let S denote the real numbers with left zero multiplication and the usual topology, and let $R = \Delta(S) \cup (N \times N)$, where N is the set of all integers. Then S is a locally compact σ-compact semigroup, R is a closed congruence on S, but S/R is not locally compact.

In 1.50, the hypothesis that S/R_1 and S/R_2 are topological semigroups can be omitted in case S is locally compact σ-compact.

As an immediate consequence of 1.56, we have a topological version of the Rees quotient semigroup.

1.57 Theorem. Let S be a locally compact σ-compact semigroup and let I be a closed ideal of S. Then S/I is a topological semigroup.

If the space of a topological semigroup S is regular and I is a closed ideal of S, then S/I is Hausdorff and multiplication is continuous at each point of $((S/I) \times (S/I))\backslash\{(\pi(I) \times (S/I)) \cup ((S/I) \times \pi(I))$. Moreover, if I is compact, then using 1.1, we obtain that multiplication is continuous at each point of $(\pi(I) \times (S/I)) \cup ((S/I) \times \pi(I))$, and hence S/I is a topological semigroup S on a metric space with a closed ideal I such that multiplication fails to be continuous on $(\pi(1) \times (S/I)) \cup ((S/I) \times \pi(I))$. We will show that multiplication is not continuous at $(\pi(I),\pi(I))$.

The following example appears in [Lawson and Madison, 1971]:

Let E denote the real line with the usual topology and let $E_2 = E \times E$ with the product topology. Define multiplication on E_2 by $((x,y),(a,b)) \to (x+a, \min(y,b))$, where + is the usual addition on E. Observe that E is a topological group under +, and let Q denote the subgroup of rational numbers. Let $S = \{(a,b) \in E_2 : a \in Q$ and $b \geq 0\}$ and $I = \{(a,b) \in S : b = 0\}$. Then S is a subsemigroup of E_2 and I is a closed ideal of S. Let $\pi : S \to S/I$ denote the natural map, and observe that any open set in S containing I is π-saturated. To demonstrate that multiplication is not continuous at $(\pi(I),\pi(I))$, we will produce an open set U in S containing I such that for any open set V in S containing I, we have that V^2 is not contained in U. To construct U, we first define a function $f : E \to E$. Let $\{a_n\}$ be a decreasing sequence of irrational numbers between 0 and 1 converging to 0. Define $f(x) = 1$ if $x \leq 0$ or x is an integer, $f(x) = 0$ if $x = n+a_n$ for $n \in \mathbb{N}$, and extend piecewise linearly for all other real numbers x. We indicate the graph of f in the following sketch:

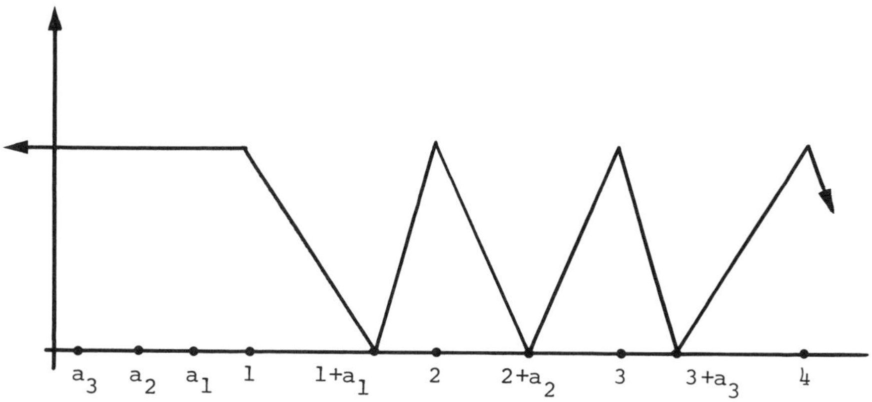

Let $U_0 = \{(a,b) \in E_2 : b < f(a)\}$. Then U_0 is open in E_2, and hence $U = U_0 \cap S$ is open in S and contains I.

Suppose that V is an open set in S containing I, and select $0 < x$ in Q and $0 < u < 1$ in E such that the intersection of the line segment L^0 joining (0,y) and (x,y) with S is contained in V. Let m be chosen in $\mathbb{N}$ and let b be chosen in E such that $a_m < x$, $0 < b < y$, and $(m,b) \in V$. Let L denote the line segment in E_2 joining (m,b) and (m + x,b). Then $L \cap S = (L_0 \cap S)(m,b)$ is contained in V^2 but not in U.

We turn now to the Monotone Light Factorization Theorem for compact semigroups. It is the semigroup analog of the classical topological result in [Eilenberg, 1934] and [Whyburn, 1934], and was proved in [Wallace, 1955]. Two topological lemmas will be employed in our proof.

Recall that if X is a space and $p \in X$, then the *quasi-component* of p in X is the intersection of all subsets of X containing p which are both open and closed. Observe that each component of X is contained in a unique quasi-component.

1.58 Lemma. Let X be a locally compact Hausdorff space and let Q be a compact quasi-component of X. Then Q is connected.

Proof. Suppose that $Q = A \cup B$, where A and B are disjoint subsets of X which are closed in Q. Since Q is compact, we have that A and B are compact. Let C be a component of X contained in Q. We can assume that $C \subset A$. Using the local compactness of X and the compactness of A, we can find an open set U such that $A \subset U \subset \overline{U} \subset X \backslash B$ and $F(U) = \overline{U} \backslash U$ (the boundary of U) is compact. For each $x \in F(U)$, there exists an open and closed set G containing x such that $G \cap Q = \square$, since $x \notin Q = \cap \{V : V \text{ is an open and closed set containing } C\}$. Since F(U) is compact, there exists a finite cover $\{G_1, \ldots, G_n\}$ of F(U) by open and closed sets such that no member meets Q. Let $W = G_1 \cup \cdots \cup G_n \cup U$. Then W is open and closed, $A \subset W$, and $W \cap B = \square$. ■

As a consequence of 1.58, we have that each component of a compact Hausdorff space is a quasi-component.

Recall that a function $f : X \to Y$ from a space X onto a space Y is *monotone* [*light*] if $f^{-1}(p)$ is connected [totally disconnected] for each $p \in Y$.

1.59 Lemma. Let X and Y be spaces and let $f : X \to Y$ be a monotone quotient map. Then $f^{-1}(C)$ is connected for each connected subset C of Y.

Proof. Let C be a connected subset of Y and let N be a nonempty open and closed subset of $f^{-1}(C)$. Since f is monotone, N is f-saturated, and hence f(N) is open and closed in C. It follows that $f(N) = C$, so $N = f^{-1}(C)$ and $f^{-1}(C)$ is connected. ∎

1.60 Monotone Light Factorization Theorem. Let S and T be compact semigroups and let $\phi : S \to T$ be a continuous surmorphism. Then there exists a compact semigroup K, a continuous monotone surmorphism $\alpha : S \to K$, and a continuous light surmorphism $\beta : K \to T$ such that the diagram:

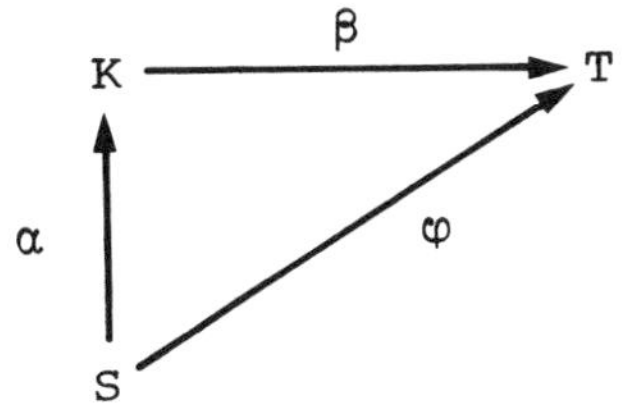

commutes.

Proof. Let R denote the set of pairs $(x,y) \in S \times S$ such that x and y are in the same component of $\phi^{-1}(t)$ for some $t \in T$. We will establish that R is a closed congruence on S. It is clear that R is an equivalence.

To see that R is a congruence, let (a,b) and (c,d) be in R. Then a and b are in the same component C_1 of $\phi^{-1}(t_1)$ for some $t_1 \in T$, and c and d are in the same component C_2 of $\phi^{-1}(t_2)$ for some $t_2 \in T$. We obtain that ac and bd are in $C_1C_2 \subset \phi^{-1}(t_1)\phi^{-1}(t_2) \subset \phi^{-1}(t_1t_2)$, and C_1C_2 is connected. It follows that $(ab,cd) \in R$, and hence R is a congruence.

To see that R is closed, let $(x,y) \in (S \times S)\backslash R$. If $\phi(x) \neq \phi(y)$, then $(x,y) \in (S \times S)\backslash K(\phi) \subset (S \times S)\backslash R$ and $(S \times S)\backslash K(\phi)$ is open. We assume then that x and y lie in different components D_1 and D_2, respectively, of $\phi^{-1}(t)$ for some $t \in T$. In view of 1.58, we see that D_1 is an intersection of open and closed (compact) subsets of $\phi^{-1}(t)$, and hence there exists P open and closed in $\phi^{-1}(1)$ such that $D_1 \subset P \subset S\backslash D_2$. It follows that P and $\phi^{-1}(t)\backslash P$ are compact and hence are disjoint closed subsets of S. Using normality, we obtain an open subset W of S such that $P \subset W \subset \overline{W}$ and $\overline{W} \cap (\phi^{-1}(t)\backslash P) = \square$. Let $U = W \phi^{-1}\phi(F(W))$, where $F(W) = \overline{W}\backslash W$, and let $V = S\backslash\overline{W}$. Then $(x,y) \in U \times V$ and U and V are open in S. To complete the argument that R is closed, we prove that $U \times V \subset (S \times S)\backslash R$.

Suppose $(U \times V) \cap R \neq \square$ and let $(a,b) \in (U \times V) \cap R$. Then a and b are in the same component of $\phi^{-1}(t_0)$ for some $t_0 \in T$. Now $a \in W$ and $b \in S\backslash\overline{W}$. Since C is connected and contains a and b, we have $C \cap F(W) \neq \square$. Let $p \in C \cap F(W)$. Then $\phi(p) = \phi(a) = t_0$ and $a \in \phi^{-1}(F(W))$. This contradicts that $a \in U = W \phi^{-1}\phi(F(W))$, and it follows that R is closed.

Let $K = S/R$ and $\alpha : S \to K$ the natural map. Since $K(\alpha) \subset K(\phi)$, we obtain the desired diagram from 1.48. It is immediate from the definition of R that α is monotone. Using 1.59, we obtain that β is light. ∎

We turn our attention to continuous homomorphisms of topological groups. The reader is referred to [Pontryagin, 1966] for a broader development of this topic.

1.61 Theorem. Let G and H be topological groups and let $\phi : G \to H$ be a continuous surmorphism. Then ϕ is quotient if and only if ϕ is open.

Proof. If ϕ is open, then clearly ϕ is quotient. Suppose, on the other hand, that ϕ is quotient, and let U be an open subset of G. Denote by e the identity of H. Then $\phi^{-1}\phi(U) = \phi^{-1}(e)U = \bigcup_{x\in\phi^{-1}(e)} \{xU\}$ is open, since translation in G is a homomorphism. Since ϕ is quotient, we have that $\phi(U)$ is open, and hence ϕ is open. ∎

1.62 Theorem. Let G be a locally compact σ-compact group, H a locally compact group, and $\phi : G \to H$ a continuous surmorphism. Then ϕ is open.

Proof. Let e and f denote the identities on G and H, respectively. It suffices to show that for each open set U in G containing e, there exists an open set W in H containing f such that $W \subset \phi(U)$. Let U be an open set in G containing e and let V be an open set in G containing e such that $F = \overline{V}$ is compact and $FF^{-1} \subset U$. Let $\{K_j\}$ be a sequence of compact subsets of G which cover G. Now $\{Vx : x \in K_j\}$ is an open cover of K_j for each $j \in \mathbb{N}$. Since each K_j is compact, we obtain a sequence $\{a_n\}$ in G such that $\{Fa_n : n \in \mathbb{N}\}$ covers G. Let $F_n = Fa_n$ for each $n \in \mathbb{N}$. We claim that $\phi(F)$ contains an open set M in H. Otherwise $\phi(F_n)$ does not contain an open set in H for each $n \in \mathbb{N}$, contrary to Baire's Category Theorem.

Fix $x \in M$ and choose $a \in F$ such that $\phi(a) = x$. Since $FF^{-1} \subset U$, we have that $Fa^{-1} \subset U$, and hence $Mx^{-1} \subset \phi(F)x^{-1} = \phi(F)\phi(a^{-1}) = \phi(Fa^{-1}) \subset \phi(U)$. Since Mx^{-1} is open and contains f, we have the desired conclusion. ■

In the paragraph preceding 1.50 an example is given of a continuous surmorphism from a locally compact σ-compact semigroup S onto a compact semigroup T which is not quotient. In view of this example, we see that 1.62 does not generalize to topological semigroups.

Let $\mathbb{R}$ denote the additive group of all real numbers with the usual topology and let $\mathbb{Z}$ denote the subgroup of integers. Let $R = \{(x,y) \in \mathbb{N} \times \mathbb{R} : y = x + n \text{ for some } n \in \mathbb{Z}\}$. Then R is a closed congruence on the locally compact σ-compact group $\mathbb{R}$. The quotient group $\mathbb{R}/R$ is usually denoted by $\mathbb{R}/\mathbb{Z}$. We will show that $\mathbb{R}/\mathbb{Z}$ is topologically isomorphic to the circle group T. Define $\phi : \mathbb{R} \to T$ by $\phi(x) = \exp(2\pi i x)$. Then ϕ is a continuous surmorphism, and hence by 1.62, ϕ is quotient. Observing that $K(\phi) = R$ and using 1.49, we have that $\mathbb{R}/\mathbb{Z}$ is topologically isomorphic to T.

The preceding example illustrates the proposition that a closed congruence on a topological group is determined by a closed normal subgroup. The converse is also true. If H is a closed normal subgroup of a topological group G, then $R = \{(x,y) \in G \times G : xH = yH\}$ is a closed congruence on G. On the other hand, if R is a closed congruence on G, then $H = \{x \in G : (x,e) \in R\}$, where e is the identity of G, is a closed normal subgroup of G and R is determined by H as above. As in the example above, the quotient group G/R is usually denoted by G/H. Unlike the situation for topological semigroups in general, the quotient group G/H is a topological group without the assumption of any additional hypothesis as in 1.56. To see this, let $\pi : G \to G/H$ denote the natural map and let U be an open subset of G. Then $\pi^{-1}\pi(U) = UH$ is open, so that $\pi(U)$ is open, since π is quotient. It follows that π is open and hence $\pi \times \pi$ is open. In view of 1.51, we see that G/H is a topological semigroup. The continuity of inversion in G/H follows from the fact that π is open and from the commutativity of the diagram:

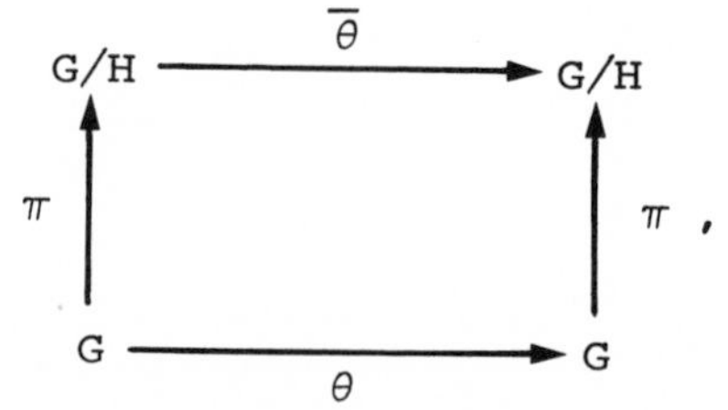

where θ and $\overline{\theta}$ are inversions in G and G/H, respectively.

If S and T are semigroups, we write $S \approx T$ to mean that S is algebraically isomorphic to T, and $S \hookrightarrow T$ to mean that S is algebraically isomorphic to a subsemigroup of T. In the case that S and T are topological semigroups, we write $S \overset{\tau}{\approx} T$ to mean that S is topologically isomorphic to T, and $S \overset{\tau}{\hookrightarrow} T$ to mean that S is topologically isomorphic to a subsemigroup of T.

We conclude this section with some examples dealing with homomorphisms and congruences.

Let $I = [0,\frac{1}{2}]$ in the usual interval I_u. Then I is a closed ideal of I_u. Define $\phi : I_u \to I_n$ by

$$\phi(x) = \begin{cases} x & \text{if } \frac{1}{2} \leqslant x; \\ \frac{1}{2} & \text{if } x > \frac{1}{2}. \end{cases}$$

Then ϕ is a continuous surmorphism such that $K(\phi) = K(\pi)$, where $\pi : I_u \to I_u/I$ is the natural map. In view of 1.49, we see that $I_n \overset{\tau}{\approx} I_u/I$.

Let $r \in \mathbb{H}^*$ and let $I = [r,\infty]$ in $\mathbb{H}^*$. Then I is a closed ideal of $\mathbb{H}^*$. Let $\pi : \mathbb{H}^* \to \mathbb{H}^*/I$ denote the natural map. We will use $\mathbb{H}^*_r$ to denote the semigroup $\mathbb{H}^*/I$ and for $0 < r$ $\mathbb{H}_r$ to denote the subset $[0,r)$ of $\mathbb{H}^*$. It is a simple exercise to show, for $0 < r < \infty$, that $\mathbb{H}^*_r \overset{\tau}{\approx} I_n$ and that $\mathbb{H}^*_\infty \overset{\tau}{\approx} \mathbb{H}^*$. (Consider the map $f : \mathbb{H}^* \to I_u$ defined by $f(x) = (1/2)^{x/r}$ for $0 \leqslant x < \infty$ and $f(\infty) = 0$, and apply the First Isomorphism Theorem (1.49)).

The compact semigroup $\mathbb{H}^*_r$, for $0 < r < \infty$, can also be obtained as a quotient of the locally compact σ-compact semigroup $\mathbb{H}$.

Let $0 < r < \infty$ and let $I = [r,\infty)$. Then I is a closed ideal of $\mathbb{H}$. Let $\pi : \mathbb{H} \to \mathbb{H}/I$ be the natural map. Then $\pi([0,r]) = \mathbb{H}/I$, and hence $\mathbb{H}/I$ is compact. It is a simple exercise to show that $\mathbb{H}^*_r \overset{\tau}{\approx} \mathbb{H}/I$.

Let $R = \{(x,y) \in \mathbb{D} \times \mathbb{D} : |x| = |y|\}$. Then R is a closed congruence on $\mathbb{D}$ and hence $\mathbb{D}/R$ is a compact semigroup. Observe that the map $\phi : \mathbb{D} \to I_u$ defined by $\phi(x) = |x|$ is a continuous surmorphism of $\mathbb{D}$ onto I_u, and that $R = K(\phi)$. It follows from 1.49 that $I_u \overset{\tau}{\approx} \mathbb{D}/R$.

Chapter 2

NEW SEMIGROUPS FROM OLD

In the study of most mathematical disciplines an attempt is made to form new (and usually more complicated) objects from a given object or collection of objects. This chapter is devoted to such considerations. Among the techniques are the formation of a variety of types of products, adjunction semigroups, projective and injective limits, semigroups of homomorphisms, compactifications, and free semigroups.

CARTESIAN PRODUCTS

As in the case of topological spaces, one of the fundamental methods of obtaining a new topological semigroup from a given collection of topological semigroups is to form their cartesian product. In this section we discuss this technique along with some of the properties that the cartesian product inherits from the original collection.

If $\{X_i\}_{i\in I}$ is a collection of sets [spaces] indexed by a nonempty set I, then we use $\Pi\{X_i\}_{i\in I}$ to denote the cartesian product [with the product topology] and $\pi_j : \Pi\{X_i\}_{i\in I} \to X_j$ to denote the jth projection for each $j \in I$.

Recall that if $\{X_i\}_{i\in I}$ is a collection of sets [spaces], then $\Pi\{X_i\}_{i\in I}$ is the set [space] of all functions $f : I \to \cup \{X_i\}_{i\in I}$ such that $f(j) \in X_j$ for each $j \in I$, and $\pi_j(f) = f(j)$ for each $j \in I$ and each $f \in \Pi\{X_i\}_{i\in I}$. Moreover, in the topological setting, the product topology is the coarsest topology relative to which each π_j is continuous.

Let $\{S_i\}_{i\in I}$ be a collection of [topological] semigroups. Then coordinatewise multiplication on $\Pi\{S_i\}_{i\in I}$ is given by

$(fg)(j) = f(j)g(j)$, the latter product being taken in S_j for each $j \in I$.

2.1 Theorem. Let $\{S_i\}_{i \in I}$ be a collection of [topological] semigroups and $S = \Pi\{S_i\}_{i \in I}$. Then S with coordinatewise multiplication is a [topological] semigroup and each projection $\pi_j : S \to S_j$ is a [continuous open] surmorphism.

Proof. Observe that $S \neq \square$ by virtue of the axiom of choice. The associativity of multiplication on S follows from that of the semigroups in the collection $\{S_i\}_{i \in I}$. It is well known, and easy to show, that π_j is a [continuous open] surjection for each $j \in I$ and the homomorphism property is an immediate consequence of the definition of multiplication on S. Finally, in the topological semigroup case, the continuity of multiplication on S follows from the fact that its composition with each projection is continuous. ■

If $\{S_i\}_{i \in I}$ is a collection of [topological] semigroups and $S = \Pi[S_i\}_{i \in I}$, then among the properties inherited by S from the collection are: abelian, group, compact, connected, and topological group.

Some of the internal structure of [topological] semigroups is preserved by forming cartesian products. Along this line we have:

2.2 Theorem. Let $\{S_i\}_{i \in I}$ be a collection of semigroups and let $S = \Pi\{S_i\}_{i \in I}$. Then:

(a) If $e_i \in S_i$ for each $i \in I$ and $e \in S$ is defined by $e(i) = e_i$ for each $i \in I$, then $e \in E(S)$ if and only if $e_i \in E(S_i)$ for each $i \in I$. If these equivalent conditions hold, then $H(e) = \Pi\{H(e_i)\}_{i \in I}$. Moreover, e is an identity for S if and only if e_i is an identity for S_i for each $i \in I$.

(b) If $A_i \subset S_i$ for each $i \in I$, then $\Pi\{A_i\}_{i \in I}$ is a [left, right] ideal of S if and only if A_i is a [left, right] ideal of S_i for each $i \in I$. Moreover, the same equivalence holds for minimal [left, right] ideals.

Proof. To prove (a) observe that each of the following are equivalent: $e \in E(S)$; $e = e^2$; $\pi_i(e^2)$ for each $i \in I$; $e_i = e_i^2$ for each $i \in I$; and $e_i \in E(S_i)$ for each $i \in I$.

For the second part of (a), assume that $e \in E(S)$. It is clear that $\Pi\{H(e_i)\}_{i\in I} \subset H(e)$, since this product is a subgroup of S containing e. The other inclusion is a consequence of the fact that surmorphisms preserve subgroups.

The final part of (a) is transparent.

To prove (b), let $A = \Pi\{A_i\}_{i\in I}$. Now if A is a left ideal, then $A_i = \pi_i(A)$ is a left ideal, since π_i is a surmorphism for each $i \in I$. On the other hand, if each A_i is a left ideal of S_i, then $A = \cap \{\pi_i^{-1}(A_i)\}_{i\in I}$ is a left ideal, since inverse images of left ideals are left ideals and likewise for non-empty intersections.

If A is minimal, then so is each A_i since surmorphic images of minimal left ideals are again minimal left ideals. On the other hand, if A_i is a minimal left ideal of S_i for each $i \in I$ and $a \in A$, then $Aa = \Pi\{A_i a(i)\}_{i\in I} = \Pi\{A_i\}_{i\in I} = A$, and we conclude that A is a minimal left ideal. ∎

If D is a directed set, then a family $\{A_\alpha\}_{\alpha\in D}$ is said to be a descending family indexed by D provided $\alpha \leqslant \beta$ in D implies $A_\beta \subset A_\alpha$. Observe that a descending family indexed by a directed set is a descending family. Since no confusion is likely, we write simply $\{A_\alpha\}_{\alpha\in D}$ is a descending family.

<u>*2.3 Theorem*</u>. Let $\mathcal{R} = \{R_\alpha\}_{\alpha\in D}$ be a descending family of [closed] congruences on a [compact] semigroup S. Let $R = \cap \{R_\alpha\}_{\alpha\in D}$ and let $\pi : S \to S/R$ and $\pi_\alpha : S \to S/R_\alpha$ denote the natural maps. For $\alpha \in D$ and $\beta \leqslant \gamma$ in D, let ϕ_α and ϕ_β^γ be the unique [continuous] surmorphism making the following diagrams commute (1.48):

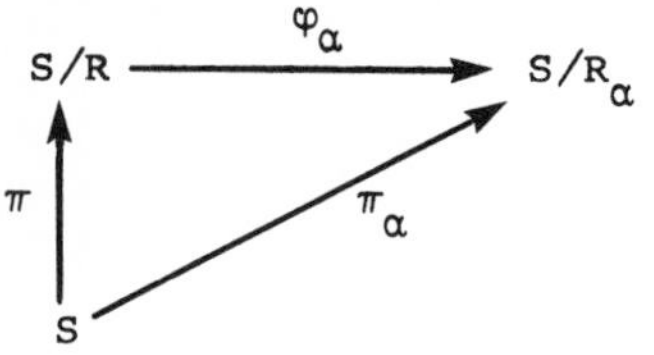

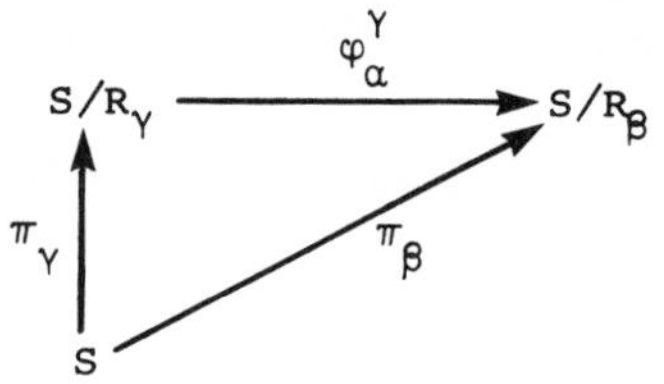

Then there is a unique [continuous] homomorphism ψ such that the diagram:

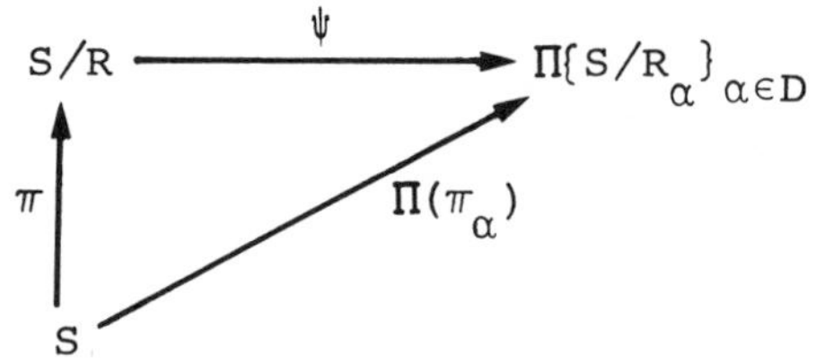

commutes, where $\Pi(\pi_\alpha)$ is the evaluation map defined by $\Pi(\pi_\alpha)(s)(\beta) = \pi_\beta(s)$ for all $s \in s$ and $\beta \in D$, and ψ is a [topological] isomorphism into. Hence, S/R is [topologically] isomorphic to a [closed] subsemigroup of $\Pi\{S/R_\alpha\}_{\alpha\in D}$.

Proof. It is clear that $R = K(\pi) = K(\Pi(\pi_\alpha))$ and the conclusion follows from 1.48 and 1.49. ■

If S,T is a pair of [topological] semigroups, we denote their cartesian product by $S \times T$.

We consider now some examples of semigroups resulting from the formation of cartesian products.

Let $S_n = I_u$ (the usual interval) for each $n \in \mathbb{N}$ and let $S = \Pi\{S_n\}_{n\in\mathbb{N}}$. Then S is a compact connected abelian semigroup with zero and identity whose underlying space is the Hilbert cube.

Let $G_n = \{-1,1\}$ (the multiplicative two point discrete group) for each $n \in \mathbb{N}$. Then $G = \Pi\{G_n\}_{n\in\mathbb{N}}$ is a topological group whose underlying space is the Cantor set. This yields a proof that the Cantor set is homogeneous.

Many interesting semigroups may be obtained by taking homomorphic images of subsemigroups of product semigroups. In some cases it is possible to realize a set with a given multiplication as a subsemigroup or as a quotient of a product semigroup, thus alleviating the need to verify associativity (or continuity) of multiplication. This can be particularly helpful when looking for examples. The next five examples illustrate these techniques.

Consider the compact semigroup $(\mathbb{R}/\mathbb{Z}) \times I_u$ on a cylinder. Observe that the minimal ideal is $(\mathbb{R}/\mathbb{Z}) \times \{0\}$. We can obtain the unit complex disk $\mathbb{D}$ as a surmorphic image of $(\mathbb{R}/\mathbb{Z}) \times I_u$. Define $\phi : (\mathbb{R}/\mathbb{Z}) \times I_u \to \mathbb{D}$ by $\phi(z,r) = zr$ (considering $\mathbb{R}/\mathbb{Z}$ and I_u as subsemigroups of $\mathbb{D}$). Then $K(\phi) = ((\mathbb{R}/\mathbb{Z}) \times \{0\}) \cup ((\mathbb{R}/\mathbb{Z}) \times \{0\}) \Delta((\mathbb{R}/\mathbb{Z}) \times I_u)$, and hence by 1.49, we have that $((\mathbb{R}/\mathbb{Z}) \times I_u)/((\mathbb{R}/\mathbb{Z}) \times \{0\}) \overset{\tau}{\approx} \mathbb{D}$.

Let S be a locally compact σ-compact semigroup and $T = S \times I$, where I represents any one of I_u, I_n, or I_m. Then $S \times \{0\}$ is a closed ideal of T (where 0 denotes 1/2 in the case of I_n) and the Rees quotient $T/(S \times \{0\})$ is a topological semigroup whose underlying space is the cone over S. This technique can be used to show that each locally compact σ-compact semigroup is embeddable in a connected monoid.

Define a multiplication on the set $S = \mathbb{D} \cup \{a,1\}$ by letting 1 be an identity for all of S, $a^2 = (\frac{1}{2},0) \in \mathbb{D}$, $a(x,y) = (x,y)a = (\frac{\sqrt{2}}{2},0)(x,y)$ for all $(x,y) \in \mathbb{D}$, and multiplication by usual multiplication on $\mathbb{D}$. If one wants to show that this multiplication is associative, it can be observed that S is isomorphic to the following subsemigroup T of $\mathbb{D} \times I_m$:

$$T = (\mathbb{D} \times \{\tfrac{1}{2}\}) \cup \{((\tfrac{\sqrt{2}}{2},0),\tfrac{\sqrt{2}}{2}),((1,0),1)\}.$$

The previous example can be thought of as taking a particular semigroup S and "pulling up" a part of S both topologically and algebraically. The technique of "pulling up" intervals can be quite useful. Suppose we would like to pull up the interval $[\frac{1}{2},1]$ from the multiplicative semigroup $[-1,1]$. Then we simply consider the subsemigroup T of $[-1,1] \times I_n$ defined by $T = ([-1,1] \times \{\frac{1}{2}\}) \cup \{(x,x) : \frac{1}{2} \leqslant x \leqslant 1\}$. The underlying space of T is illustrated as follows:

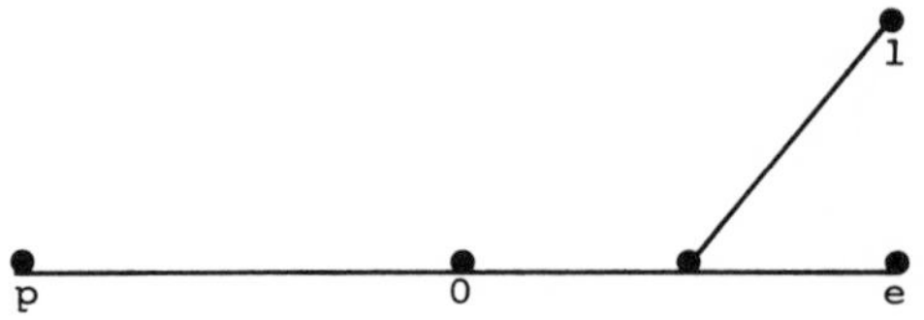

Notice that $E(T) = \{(0,\frac{1}{2}),(1,\frac{1}{2}),(1,1)\}$, denoted on the figure 0, e, and 1, respectively. It is easy to check that $H(1) = \{1\}$ and $H(e) = \{e,p\}$. This simple example has certain pathological properties and will be referred to later in this treatise.

Let $\eta : \mathbb{R} \to \mathbb{R}/\mathbb{Z}$ denote the natural map and $S = \{(\eta(x),x) : x \in \mathbb{H}\} \cup \{(\mathbb{R}/\mathbb{Z}) \times \{\infty\}\}$. Then S is a compact connected subsemigroup of $(\mathbb{R}/\mathbb{Z}) \times \mathbb{H}^*$ containing the identity (1,0) and the minimal ideal $(\mathbb{R}/\mathbb{Z}) \times \{\infty\}$. The underlying space of S is a circle with a ray spiralling down on the circle as illustrated below:

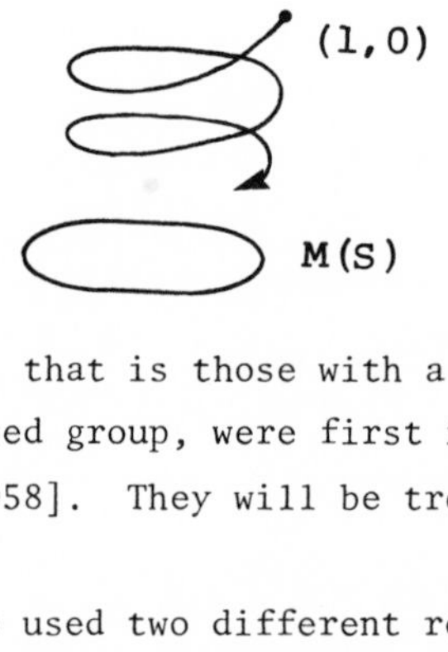

Semigroups of this type, that is those with a copy of $\mathbb{H}$ winding down toward a compact connected group, were first isolated and described in [Koch and Wallace, 1958]. They will be treated in greater detail in Chapter 5.

Notice that we have used two different representations of the same semigroup on the cylinder. We will consistently use that representation of a given semigroup which appears to best lend itself to the given situation.

One final type of subsemigroup of product semigroups will be of use to us and is most likely familiar to the reader in the algebraic setting. This is the direct sum of a family of [topological] monoids.

Let $\{S_i\}_{i\in I}$ be a family of [topological] monoids (with $I \neq \square$) and let $S = \Pi\{S_i\}_{i\in I}$. Then the *direct sum* of the family $\{S_i\}_{i\in I}$ is the subset T of S consisting of those members f of S satisfying $f(i) = 1_i$ (the identity of S_i), for all but finitely many $i \in I$. The set T is denoted by $\Sigma\{S_i\}_{i\in I}$.

<u>*2.4 Theorem*</u>. Let $\{S_i\}_{i\in I}$ be a family of [topological] monoids. Then:

(a) $\Sigma\{S_i\}_{i\in I}$ is a [dense] subsemigroup of $\Pi\{S_i\}_{i\in I}$;

(b) $\Sigma\{S_i\}_{i\in I}$ is abelian if and only if S_i is abelian for each $i \in I$; and

(c) $\Sigma\{S_i\}_{i\in I}$ is a [topological] group if and only if S_i is a [topological] group for each $i \in I$.

SEMIDIRECT PRODUCTS

In this section we discuss semidirect products of semigroups, and illustrate this concept with some examples. The equivalence of this notion with split exact sequences for groups is demonstrated. In order to present semidirect products in their topological setting we begin with some topological preliminaries.

If X and Y are Hausdorff spaces, then C(X,Y) denotes the set of all continuous functions from X into Y. For $A \subset X$ and $B \subset Y$, we use N(A,B) to denote $\{f \in C(X,Y) : f(A) \subset B\}$. The topology on C(X,Y) having the collection of all N(K,U) such that K is a compact subset of X and U is an open subset of Y as a subbase is called the *compact-open topology* on C(X,Y).

<u>*2.5 Lemma*</u>. Let X and Y be Hausdorff spaces. Then C(X,Y), with the compact-open topology, is a Hausdorff space.

Proof. Suppose that f and g are distinct elements of C(X,Y). Then there exists $x \in X$ such that $f(x) \neq g(x)$. Let U and V be disjoint open subsets of Y containing f(x) and g(x), respectively. We have that $f \in N(\{x\},U)$, $g \in N(\{x\},V)$, and $N(\{x\},U) \cap N(\{x\},V) = \square$, and we conclude that C(X,Y) is Hausdorff. ∎

<u>*2.6 Lemma*</u>. Let X be a locally compact Hausdorff space. Then C(X,X) is a topological semigroup relative to the compact-open topology and composition of functions.

Proof. Let K be a compact subset of X, U an open subset of X, and suppose that f and g are in C(X,X) such that $fg \in N(K,U)$. Then g(K) is a compact subset of the open set $f^{-1}(U)$. Since X is locally compact, there exists an open set V such that $\overline{V}$ is compact and $g(K) \subset V \subset \overline{V} \subset f^{-1}(U)$. We now have $f \in N(\overline{V},U)$, $g \in N(K,V)$, and $N(\overline{V},U)N(K,V) \subset N(K,U)$. It follows that multiplication in C(X,X) is continuous. The associativity is transparent. ■

For a locally compact Hausdorff space X, we will assume for the remainder of the section that C(X,X) is assigned the compact-open topology.

<u>*2.7 Lemma*</u>. Let X be a locally compact Hausdorff space. Then the map $C(X,X) \times X \to X$ defined by $(f,x) \to f(x)$ is continuous.

Proof. Let $(f,x) \in C(X,X) \times X$ and let U be an open subset of X containing f(x). Then $f^{-1}(U)$ is an open subset of X containing x. Let V be an open set such that $\overline{V}$ is compact and $x \in V \subset \overline{V} \subset f^{-1}(U)$. Then $(f,x) \in N(\overline{V},U) \times V$ and if $(g,a) \in N(\overline{V},U) \times V$, then $g(a) \in g(v) \subset g(\overline{V}) \subset U$. ■

The map of 2.7 is frequently referred to as the *evaluation map*.

If S is a [topological] semigroup, then we use End(S) to denote the set of [continuous] endomorphisms of S (homomorphisms of S into itself).

<u>*2.8 Theorem*</u>. Let S be a [locally compact] semigroup. Then End(S) [with the relative topology of C(S,S)] is a [topological] semigroup under composition of [continuous] homomorphisms.

Proof. Composition of homomorphisms is clearly associative, and the topological semigroup case is an immediate consequence of 2.6. ■

In the topological setting we have as a consequence of 2.7 that the evaluation map is continuous:

2.9 Theorem. Let S be a locally compact semigroup. Then the map $\mathrm{End}(S) \times S \to S$ defined by $(f,x) \to f(x)$ is continuous.

If S is a [locally compact] semigroup and T is a [topological] semigroup acting as a semigroup of [continuous] endomorphisms of S, i.e., there exists a [continuous homomorphism $\phi : T \to \mathrm{End}(S)$, then we define the semidirect product of S and T to be $S \times T$ [with the product topology] together with multiplication

$$((s_1,t_1),(s_2,t_2)) \to (s_1\phi(t_1)(s_2),t_1t_2)$$

We denote the semidirect product by $S \times_\phi T$, where $t \in T$ and $s \in S$. The semidirect product of S and T is also called the *split extension of* S *by* T and denoted $\mathrm{Split}(S,T,\phi)$.

2.10 Theorem. Let S be a [locally compact] semigroup, T a [topological] semigroup, and $\phi : T \to \mathrm{End}(S)$ a [continuous] homomorphism. Then $S \times_\phi T$ is a [topological] semigroup.

Proof. The proof that multiplication is associative on the semidirect product is tedious but straightforward. To see that multiplication is continuous in the topological semigroup case we adopt the following notation:

(i) m is the multiplication on $S \times_\phi T$;

(ii) m_S and m_T are the multiplications on S and T, respectively;

(iii) $\pi_S : S \times T \to S$ and $\pi_T : S \times T \to T$ are projections;

(iv) $\pi_{2,4} : (S \times T) \times (S \times T) \to T \times T$ is projection onto the second and fourth factor;

(v) $\pi_{1,2,3} : (S \times T) \times (S \times T) \to S \times T \times S$ is projection onto the first, second, and third factor;

(vi) 1_S is the identity map on S; and

(vii) $e : \mathrm{End}(S) \times S \to S$ is evaluation.

To show that m is continuous, it suffices to show that $\pi_S m$ and $\pi_T m$ are continuous. The continuity of these maps follows from the observation that $\pi_S m = m_S(1_S \times e)(1_S \times \phi \times 1_S)\pi_{1,2,3}$ and $\pi_T m = m_T \pi_{2,4}$. ∎

We now consider some examples of semidirect products.

Let $S = \mathbb{R}$ and let T be the semigroup of multiplicative reals with the usual topology. Define $\phi : T \to \mathrm{End}(\mathbb{R})$ by $\phi(t)(x) = tx$ (ordinary multiplication of real numbers). Then multiplication on $\mathbb{R} \times_\phi T$ is given by $(x,s)(y,t) = (x + sy, st)$. Without resorting to the continuity of ϕ we see that this is a continuous multiplication relative to the product topology on $\mathbb{R} \times T$. Let A denote the subsemigroup of $M(2,\mathbb{R})$ of matrices of the form $\begin{bmatrix} a & b \\ 0 & 1 \end{bmatrix}$, for $a, b \in \mathbb{R}$, and define $\psi : \mathbb{R} \times_\phi T \to A$ by $\psi(x,s) = \begin{bmatrix} s & x \\ 0 & 1 \end{bmatrix}$. Then it is readily verified that ψ is a topological isomorphism of $\mathbb{R} \times_\phi T$ onto A. Let us also define δ from $\mathbb{R} \times_\phi T$ into the set of transformations from $\mathbb{R}$ into $\mathbb{R}$ by $\delta(x,s)(y) = x + sy$. Then for each $(x,s) \in \mathbb{R} \times_\phi T$, we have that $\delta(x,s)$ is an affine transformation of the line ($f : \mathbb{R} \to \mathbb{R}$ is affine provided $f(\lambda x + (1 - \lambda)y) = \lambda f(x) + (1 - \lambda)f(y)$ for each $\lambda, x, y \in \mathbb{R}$). Moreover, δ is an algebraic isomorphism from $\mathbb{R} \times_\phi T$ onto $\mathrm{Aff}(\mathbb{R})$ (the semigroup of affine transformations of the line under composition). One sees that points above the x-axis represent orientation preserving affine homeomorphisms, points below the x-axis represent orientation reversing affine homeomorphisms, points on the x-axis represent constant transformations, and points on the y-axis represent linear transformations. The complement of the x-axis is the classical group of affine transformations of the line (see [Weyl, 1946]).

The semigroup $\mathbb{R} \times_\phi T$ has some interesting subsemigroups. These include $\{(x,s) : 0 \leqslant x,\ s \leqslant 1, \text{ and } x + s \leqslant 1\}$ which is topologically isomorphic to the triangle under the map $(x,s) \to \begin{bmatrix} s & x \\ 0 & 1 \end{bmatrix}$. Another is $B = \{x,s) : 0 \leqslant x \text{ and } 0 < s\}$ which is a locally compact semigroup with identity $(0,1)$ such that $H((0,1)) = \{(x,s) : x = 0 \text{ and } 0 < s\}$ and $M(B) = B \backslash H((0,1))$. Observe that $M(B)$ is dense in B and is therefore not a closed subspace of B. In the compact and discrete cases, the minimal ideal is closed. Until 1972 it was unknown in general whether the minimal ideal of a topological semigroup must be closed (see [Dobbins, 1972]). Additional discussion of this topic appears in [Cohen, 1973].

The previous example illustrates a useful technique of finding different realizations of the same semigroup. An application of this technique could be employed to define a topology on Aff($\mathbb{R}$) making it into a topological semigroup under composition. One does not ordinarily easily obtain a topology on a semigroup of transformations of a space which makes it into a topological semigroup. Observe that one can form the semidirect product of the additive group $(R,+)$ of any ring with its multiplicative semigroup $(R,\cdot)$ in exactly the same way as in 2.10, i.e., define $\phi : (R,\cdot) \to \mathrm{End}(R,+)$ by $\phi(r)(s) = rs$.

Let $S = [0,1]$ with the usual topology and multiplication $(x,y) \to x \wedge y = \min(x,y)$, i.e., $S = I_m$ (the min interval), and let $T = [0,1]$ with the usual topology and multiplication $(x,y) \to x \vee y = \min(x,y)$. Then S and T are compact semigroups. Define $\phi : T \to \mathrm{End}(S)$ by $\phi(t)(s) = \max(t,s)$, where $t \in T$ and $s \in S$. It is readily seen that ϕ is a continuous homomorphism. The multiplication on $S \times_\phi T$ is given by $((a,b),(x,y)) \to (a \wedge (b \wedge x), b \wedge y)$. Note that each element of $S \times_\phi T$ is idempotent and that $(1,0)$ is an identity for $S \times_\phi T$. Rectangles of the form $\{(x,y) : y \leqslant k \leqslant x\}$ are abelian subsemigroups and form a subbasis for neighborhoods of $(1,0)$.

Let $S = I_m$ and $T = I_u$. Define $\phi : T \to \mathrm{End}(S)$ by $\phi(t)(s) = ts$ (ordinary multiplication). Then ϕ is a continuous homomorphism. The element $(0,0)$ is a zero for $S \times_\phi T$, $(1,1)$ is a left identity, and there is no right identity. For each $c \in [0,1]$, $\{(cs,x) : 0 \leqslant x \leqslant 1\} \cup \{(x,1) : c \leqslant x\}$ is an abelian subsemigroup whose underlying space is an interval, and it has a zero and an identity. In fact, each such subsemigroup may be thought of as a usual interval with a min interval "sitting on top".

As we have seen, forming semidirect products is a useful technique for building a new semigroup from a given pair of semigroups. Observe that the new semigroup need not be abelian even if both of the original semigroups are abelian. The three preceding examples illustrate this fact. Several additional examples of semidirect products (split extensions) are to be found in [Hofmann and Mostert, 1966].

We turn now to a discussion of semidirect products of groups and the equivalence of this notion with that of split exact sequences.

If S is a [locally compact] semigroup, then Aut(S) denotes the group of [bicontinuous] automorphisms of S ([topological] isomorphisms of S onto itself).

If S is a [locally compact] semigroup, then Aut(S) is the group of units of End(S). If G is a [topological] group and $\phi : G \to \text{End}(S)$ is a [continuous] homomorphism, then the condition that $\phi(e) = 1_S$, where e is the identity of G, is equivalent to the condition that $\phi(G) \subset \text{Aut}(s)$. This condition will be imposed by stating that $\phi : G \to \text{Aut}(S)$ is a [continuous] homomorphism.

2.11 Theorem. Let G and H be [locally compact] groups and $\phi : G \to \text{Aut}(H)$ a [continuous] homomorphism. Then $H \times_\phi G$ is a [topological] group.

Proof. It is straightforward to show that $H \times_\phi G$ is a group in the algebraic case. The proof in the topological setting follows from 2.10. ■

A [topological] group B is said to be an *extension* of a [topological] group A by a [topological] group C if there exist [continuous] homomorphisms $\alpha : A \to B$ and $\beta : B \to C$ such that the sequence

$$0 \longrightarrow A \xrightarrow{\alpha} B \xrightarrow{\beta} C \longrightarrow 0$$

is exact, i.e., the image of an incoming homomorphism is the kernel of the outgoing homomorphism. If, in addition, there is a [continuous] homomorphism $\gamma : C \to B$ such that $\beta\gamma$ is the identity map on C, then B is called a *split extension* of A by C and

$$0 \longrightarrow A \xrightarrow{\alpha} B \underset{\gamma}{\overset{\beta}{\rightleftarrows}} C \longrightarrow 0$$

is called a *split exact sequence*.

Let G and H be [topological] groups. Then $G \times \{1\}$ is a [closed] normal subgroup of $G \times H$ and the quotient group $G \times H/G \times \{1\}$ is [topologically] isomorphic to H. Define $\alpha : G \to G \times H$ by $\alpha(q) = (2,1)$, $\beta : G \times H \to H$ by $\beta(b,h) = h$, and $\gamma : H \to G \times H$ by $\gamma(h) = (1,h)$. Then

$$0 \longrightarrow G \xrightarrow{\alpha} G \times H \underset{\gamma}{\overset{\beta}{\rightleftarrows}} H \longrightarrow 0$$

is a split exact sequence.

In the next result we restrict our attention to the algebraic and compact settings and refer the reader to [Hofmann and Mostert, 1966] for a treatment of extensions of general topological groups.

2.12 Theorem. Let $0 \longrightarrow A \xrightarrow{\alpha} B \underset{\gamma}{\overset{\beta}{\rightleftarrows}} C \longrightarrow 0$ be a split exact sequence of [compact] groups and define $\phi : C \to \mathrm{Aut}(A)$ by $\phi(c)(a) = a^{-1}(\gamma(c)\alpha(a)\gamma(c)^{-1})$. Then ϕ is a [continuous] homomorphism, $A \times_\phi C$ is a [compact] group, and $\psi : A \times_\phi C \to B$ defined by $\psi(a,c) = \alpha(a)\gamma(c)$ is a [topological] isomorphism.

Proof. Since α and γ are injective, it clearly suffices to assume that A and C are [compact] subgroups of B. Notice that A is a normal subgroup of B. With these assumptions $\phi(c)(a) = cac^{-1}$ and $\psi(a,c) = ac$, these products being taken in B. Now, $\phi(c)$ is the restriction of an inner automorphism of B to the normal subgroup A, and hence it is an automorphism of A for each $c \in C$. That ϕ is a [continuous] homomorphism into Aut(A) is straightforward. For $(a,c),(b,d) \in A \times_\phi C$, we have $\psi((a,c)(b,d) = \psi(a\phi(c)(b),cd) = a\phi(c)(d)cd = a(cbc^{-1})cd = acbd = \psi(a,c)\psi(b,d)$, and hence ψ is a homomorphism from the group $A \times_\phi C$ into the group B. The continuity of ψ is in the compact case is readily verified. It remains to show that ψ is bijective to insure that ψ is an [topological] isomorphism and, consequently, that $A \times_\phi C$ is a compact group in the topological case.

To see that ψ is injective, suppose that $\psi(a,c) = \psi(b,d)$. Then $ac = bd$ and $b^{-1}a = dc^{-1} \in A \cap C = \{1\}$. Thus, $a = b$ and $c = d$, and ψ is injective.

To see that ψ is surjective, let $b \in B$. Then $b = a\beta(b)$ for some $a \in A$, since b and $\beta(b)$ lie in the same coset of B modulo A. It follows that $b = \psi(a,\beta(b))$. ∎

Observe that under the hypothesis of 2.12 in the compact case, one could obtain that $A \times_\phi C$ is a compact group without considering continuous automorphisms of A and without taking into account the continuity of ϕ.

2.13 Theorem. Let A and C be [compact] groups and let $\phi : C \to \mathrm{Aut}(A)$ be a [continuous] homomorphism. Define $\alpha : A \to A \times_\phi C$ by $\alpha(a) = (a,1)$, $\beta : A \times_\phi C \to C$ by $\beta(a,c) = c$, and $\gamma : C \to A \times_\phi C$ by $\gamma(c) = (1,c)$. Then

$$0 \longrightarrow A \xrightarrow{\alpha} A \times_\phi C \underset{\gamma}{\overset{\beta}{\rightleftarrows}} C \longrightarrow 0$$

is a split exact sequence of [compact] groups and $\phi(c)(a) = \alpha^{-1}(\gamma(c)\alpha(a)\gamma(c)^{-1})$ for all $(a,c) \in A \times_\phi C$.

Proof. It is readily verified that α, β and γ are [continuous] homomorphisms and that the sequence is split exact. For $(a,c) \in A \times_\phi C$, we have $\gamma(c)\alpha(a)\gamma(c)^{-1} = (1,c)(a,1)(1,c)^{-1} = (1,c)(a,1)(\phi^{-1}(c)(1),c^{-1}) = (1\phi(c)(a),c)(1,c^{-1}) = (\phi(c)(a)\phi(c)(1),cc^{-1} = (\phi(c)(a),1)$ and hence $\phi(c)(a) = \alpha^{-1}(\gamma(c)\alpha(a)\gamma(c)^{-1})$. ∎

The results of 2.12 and 2.13 shows that each split exact sequence of [compact] groups gives rise to a semidirect product of [compact] groups and, conversely, each semidirect product gives rise to a split exact sequence. Moreover, it is not difficult to show that one process followed by the other is "essentially" the identity (that is, gives you what you started with). This is a recurring phenomenon in mathematics which has been developed in category theory as the equivalence of categories.

We conclude this section with a discussion of some of the noteworthy results of K. Krohn and J. Rhodes on the structure of finite semigroups [Krohn and Rhodes, 1965].

Suppose that S and T are semigroups. Then the set S^T of all functions from T into S is a semigroup under pointwise multiplication. Define $\phi : T \to \mathrm{End}(S^T)$ by $\phi(t)(f) = f \circ \rho_t$. Then ϕ is a homomorphism. The *wreath product* $S \wr T$ of S and T is defined to be $S^T \times_\phi T$, and hence $S \wr T = S^T \times T$ with multiplication defined by $((f,a),(g,b)) \to (fg \circ \rho_a, ab)$.

A semigroup S is said to *divide* a semigroup T provided T is a surmorphic image of some subsemigroup of S.

A semigroup S is said to be *combinatorial* provided each subgroup of S is degenerate.

For a class $\mathcal{S}$ of semigroups, we denoted by $K(\mathcal{S})$ the closure of $\mathcal{S}$ under division and semidirect products. Let U_3 denote the three-element semigroup obtained by adjoining an identity to the two-element right trivial semigroup, and let $\mathcal{G}$ denote the class of all finite nondegenerate simple groups.

It is proved in [Krohn and Rhodes, 1968] that:

(a) $K(\{U_3\})$ is the class of all finite combinatorial semigroups;

(b) $K(\mathcal{G} \cup \{U_3\})$ is the class of all finite semigroups; and

(c) Each finite semigroup S divides a semidirect product $S_1 \times_{\phi_1} S_2 \times_{\phi_2} \cdots \times_{\phi_{n-1}} S_n$ for some sequence $S_1, \ldots, S_n$ of semigroups which are alternatively combinatorial semigroups and finite simple groups which divide S.

The smallest n required in (c) is called the *complexity* of S. Wreath products can be employed to form the statement of (c).

REES PRODUCTS

In the two previous sections we have seen two distinct techniques of introducing semigroup structures on products of underlying sets of semigroups. Herein we present yet another technique which, although rather specialized, has proved to be most useful in the theory of semigroups.

If X and Y are non-empty sets, S is a semigroup, and $\sigma : Y \times X \to S$ is a function, then we denote by $[X,S,T]_\sigma$ the set $X \times S \times Y$ with multiplication defined by $((r,g,\ell),(s,h,t)) \to (r,g\sigma(\ell,s)h,t)$.

The proof of the following is straightforward and is therefore omitted:

2.14 Theorem. Let X and Y be non-empty sets, S a semigroup, and $\sigma : Y \times X \to S$ a function. Then $[X,S,Y]_\sigma$ is a semigroup. If, in addition, X and Y are Hausdorff spaces, S is a topological semigroup, and σ is continuous, then $[X,S,Y]_\sigma$ is a topological semigroup.

The semigroup $[X,S,Y]_\sigma$ of 2.14 is called the *Rees product of* S *over* X *and* Y *with sandwich function* σ. If S is a [topological] group, [X and Y are Hausdorff spaces, and σ is continuous], then $[X,S,Y]_\sigma$ is called a [*topological*] *paragroup*.

2.15 Theorem. Let $S = [X,G,Y]_\sigma$ be a paragroup. Then:

(a) $M(S) = S$;

(b) $L \subset S$ is a minimal left ideal of S if and only if $L = X \times G \times \{y\}$ for some $y \in Y$;

(c) $R \subset S$ is a minimal right ideal of S if and only if $R = \{x\} \times G \times Y$ for some $x \in X$;

(d) $H = S$, the maximal subgroups of S are precisely the sets of the form $\{x\} \times G \times \{y\}$ for some $x \in X$ and $y \in Y$, and each maximal subgroup of S is isomorphic to G.

Proof. To prove (a), let $(x,g,y) \in S$. Then $S(x,g,y)S = X \times G\sigma(Y \times \{x\}g\sigma(\{y\} \times X)G \times Y = X \times G \times Y = S$, so that $J((x,g,y)) = S$ for each $(x,g,y) \in S$, and hence $S = M(S)$.

To prove (b), first observe that $X \times G \times \{y\}$ is a left ideal for each $y \in Y$. Now if L is a left ideal of S and $(x,g,y) \in L$, then $X \times G \times \{y\} = S(x,g,y) \subset L$ and the result follows.

The proof of (c) is similar to the proof of (b).

Finally, to prove (d), vix $x \in X$ and $y \in Y$ and define $\alpha : G \to \{x\} \times G \times \{y\}$ by $\alpha(g) = (x, g\sigma(y,x)^{-1}, y)$. Then it is readily verified that α is an isomorphism of G onto $\{x\} \times G \times \{y\}$. The result now follows from the fact that $s = \cup \{\{x\} \times G \times \{y\} : x \in X$ and $y \in Y\}$. ■

In [Clifford and Preston, 1961; Sec. 2.5] a careful study of 0-minimal ideals and 0-simple semigroups is made. We will not give a detailed treatment of these subjects but, primarily for the purpose of generating examples, a brief study of Rees products over a group with zero follows.

Let G be a group and let H = (0) be a trivial group such that 0 is not a member of the underlying set of G. If $S = G \cup H$ and multiplication is defined by

$$x \circ y = \begin{cases} xy & \text{if } x, y \in G, \\ 0 & \text{otherwise,} \end{cases}$$

then $(S,\circ)$ is called a *group with zero*.

If $(S,\circ)$ is a group with zero, then it is a trivial matter to see that $(S,\circ)$ is a semigroup with zero, identity, and no other idempotents. Note that if G is nontrivial, then $S = G^0$.

Let S be a semigroup with zero 0. A [left, right] ideal [L,R] I is a *0-minimal* [*left, right*] *ideal* of S if [$L \neq \{0\}$, $R \neq \{0\}$] $I \neq \{0\}$ and the only [left, right] ideal of S which is properly contained in [L,R] I is $\{0\}$. If S itself is a 0-minimal ideal, and $S^2 \neq \{0\}$, then S is said to be *0-simple*.

The proof of the next theorem is similar to that for 2.15 and is left as an exercise.

<u>*2.16 Theorem*</u>. Let X and Y be non-void sets, $S = G \cup \{0\}$ a group with zero, and $\sigma : Y \times X \to S$ a function such that $\sigma(\{y\} \times X) \cap G \neq \square$ for each $y \in Y$ and $\sigma(Y \times \{x\}) \cap G =$ for each $x \in X$. Then $X \times \{0\} \times Y$ is an ideal of the Rees product $[X,S,Y]_\sigma$. Moreover, if $T = [X,S,Y]_\sigma/(X \times \{0\} \times Y)$ and $\pi : [X,S,Y]_\sigma \to T$ is the natural homomorphism, then:

(a) T is 0-simple;

(b) $L \subset T$ is a 0-minimal left ideal of T if and only if $L = \pi(X \times S \times \{y\})$ for some $y \in Y$;

(c) $R \subset T$ is a 0-minimal right ideal of T if and only if $R = \pi(\{x\} \times X \times Y)$ for some $x \in X$; and

(d) The maximal subgroups of T are precisely those subsets $\pi(\{x\} \times G \times \{y\})$ for which $\sigma(y,x) \in G$. If $\sigma(y,x) = 0$, then $\pi(\{x\} \times G \times \{y\})^2 = \{0\}$.

We will show in Chapter 3 that the minimal ideal of a compact semigroup is a paragroup. It is shown in [Clifford and Preston, 1961; Sec. 3.2], using a slightly different approach, that the semigroups T, as described in 2.16, are precisely the completely 0-simple semigroups (those 0-simple semigroups having 0-minimal left ideals and 0-minimal right ideals). The reader is referred to this reference for proofs.

Observe that in forming the Rees product semigroup [X,G,Y] one has taken three semigroups (considering X as a left zero semigroup and Y as a right zero semigroup) and obtained a semigroup structure on the product which differs from the cartesian product by a sort of a "twist" in the middle coordinate. Recall that in Chapter 1 we "twisted" the product in both coordinates of $\mathbb{N} \times \mathbb{N}$ to obtain the bicyclic semigroup. Little imagination is needed to see that multiplication could be twisted in one or more coordinates of a product in many different ways. For example, Rees products are special cases of semigroups of the type $[S_1,S_2,S_3]_\sigma$, where S_1, S_2, and S_3 are semigroups, $\sigma : S_3 \times S_1 \to S_2$, and multiplication is defined by $((r,g,l),(s,h,t)) \to (rs,g\ (l,s)h,lt)$. (Here the sandwich function σ must satisfy additional conditions which insure associativity of multiplication.) One rather unique way to introduce unusual multiplications on products was discovered by E. D. Tymchatyn in [Tymchatyn, 1969]. Briefly, he considered a [topological] semigroup S and a [topological] partial semigroup T containing S (not all products are necessarily defined in T) with the property that $STS \subset S$. Then he defined a sandwich function $\sigma : S \times S \to T$ and a multiplication on

$S \times T \times S$ by $((r,g,1),(s,h,t)) \to (r,g\sigma(1,s)h,t)$. This technique was used in the solution of a problem raised by Hofmann and Mostert concerning one-parameter semigroups bumping through a regular $\mathcal{D}$-class [Hofmann and Mostert, 1966].

ADJUNCTION SEMIGROUPS

There is a well-known pasting construction, due to Borsuk, as follows: Let $f : X \to Y$ be a map of spaces and let A be a subset of X. Then form the space obtained from the disjoint union of X and Y by identifying each $a \in A$ with its image in Y. In this section we present a [topological] semigroup analog of the Borsuk construction along with some examples.

If X and Y are disjoint spaces, then we give $X \cup Y$ the topology which is soherent with that of X and Y, i.e., a subset U of $X \cup Y$ is open if and only if $U \cap X$ is open in X and $U \cap Y$ is open in Y. Observe that if X and Y are both [locally compact, σ-compact] compact, then $X \cup Y$ is [locally compact, σ-compact] compact.

If S and T are disjoint [topological] semigroups and $\phi : S \to T$ is a [continuous] homomorphism, then we define multiplication on $S \cup T$ by:

$$(x,y) \longmapsto \begin{cases} m_S(x,y) & \text{if } x,\ y \in S; \\ m_T(x,y) & \text{if } x,\ y \in T; \\ & \\ m_T(\phi(x),y) & \text{if } x \in S \text{ and } y \in T; \\ m_T(x,\phi(y)) & \text{if } x \in T \text{ and } y \in S, \end{cases}$$

where m_S and m_T are the multiplication on S and T, respectively. We denote $S \cup T$ with this multiplication by $S \cup_\phi T$.

The proofs of the next two theorems are straightforward and will be omitted.

2.17 Theorem. Let S and T be disjoint [topological] semigroups and let $\phi : S \to T$ be a [continuous] homomorphism. Then $S \cup_\phi T$ is a [topological] semigroup.

2.18 Theorem. Let S and T be disjoint semigroups, $\phi : S \to T$ a homomorphism, I an ideal of S, and R the congruence on $S \cup_\phi T$ generated by $\{(x,\phi(x)) : x \in I\}$. Then the restriction of the natural map $\pi : S \cup_\phi T \to (S \cup T)/R$ to T is an algebraic embedding.

We now state and prove a topological version of 2.18.

2.19 Theorem. Let S and T be disjoint locally compact σ-compact semigroups, $\phi : S \to T$ a continuous homomorphism, I a closed ideal of S, and R the congruence on $S \cup_\phi T$ generated by $\{(x,\phi(x)) : x \in I\}$. Then R is a closed congruence on $S \cup_\phi T$, $(S \cup_\phi T)/R$ is a topological semigroup, and the restriction of the natural map $\pi : S \cup_\phi T \to (S \cup_\phi T)/R$ to T is a topological embedding of T into $(S \cup_\phi T)/R$.

Proof. Observe that $R = \Delta(S) \cup \Delta(T) \cup \{(x,\phi(x)) : x \in I\} \cup \{(\phi(x),x) : x \in I\} \cup \{(x,y) \in I \times I : \phi(x) = \phi(y)\}$. Each set in this union is closed and so R is closed. Now $(S \cup_\phi T)/R$ is a topological semigroup by virtue of 1.56. The restriction $\pi|T$ of π to T is clearly a continuous isomorphism. To see that $\pi|T$ is open, let U be an open subset of T. Then $\pi^{-1}\pi(U) \cap T = U$, since $R \cap (T \times T) = \Delta(T)$. In view of the topology on $S \cup_\phi T$ and the fact that π is quotient, we see that $\pi(U)$ is open in $\pi(T)$. ■

The [topological] semigroup $(S \cup T)/R$ in 2.19 is called the *adjunction semigroup of* S *and* T *relative to* ϕ *and* I, and is denoted $S \underset{\phi,I}{\cup} T$. Observe that $S \underset{\phi,S}{\cup} T \overset{T}{\approx} S \cup_\phi T$.

Let S be a locally compact σ-compact semigroup, I a closed ideal of S, $T = \{0\}$ a trivial semigroup, and $\phi : S \to T$ the trivial homomorphism. Then $S \underset{\phi,I}{\cup}$ is topologically isomorphic to S/I.

Let S be the semigroup consisting of a ray winding down on a circle. Let T be a copy of S which is disjoint from S and let $\phi : S \to T$ be the identification. Then $S \underset{\phi,M(S)}{\cup} T$ is a semigroup consisting of two rays winding down on the same circle.

Let $\pi : \mathbb{R} \to \mathbb{R}/\mathbb{Z}$ denote the natural map and let $\phi : \mathbb{H} \to \mathbb{R}/\mathbb{Z}$ be the restriction of π to $\mathbb{H}$. For $x \in \mathbb{H}$, let $I = J(x)$, and let $e \in \pi(1)$. We illustrate the compact semigroup $\mathbb{H} \underset{\phi,I}{\cup} \mathbb{R}/\mathbb{Z}$ in the following diagrams for various choices of x.

In case x = 1, we have:

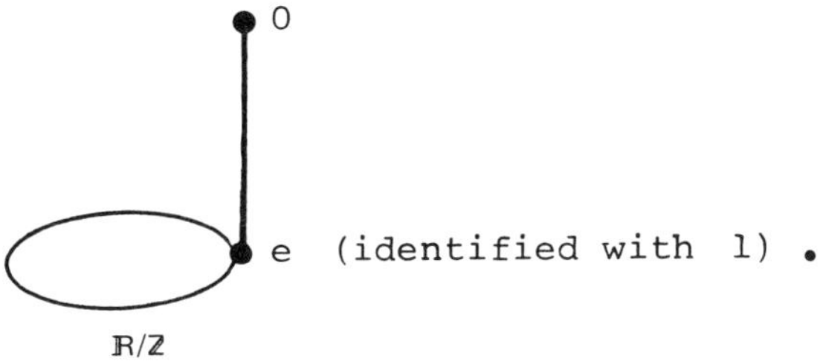

In case x = 1/2, we have:

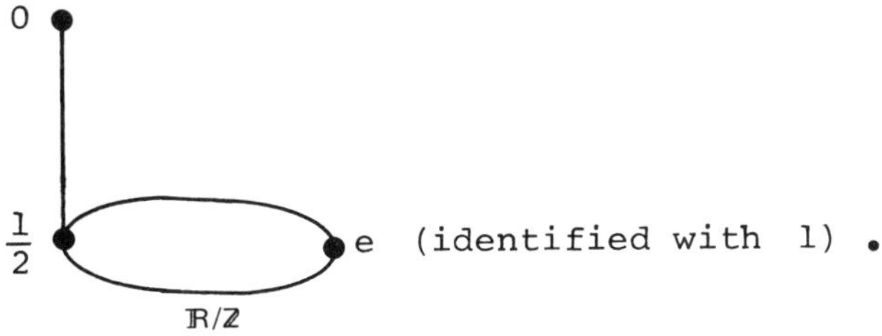

In case x = 2, we have:

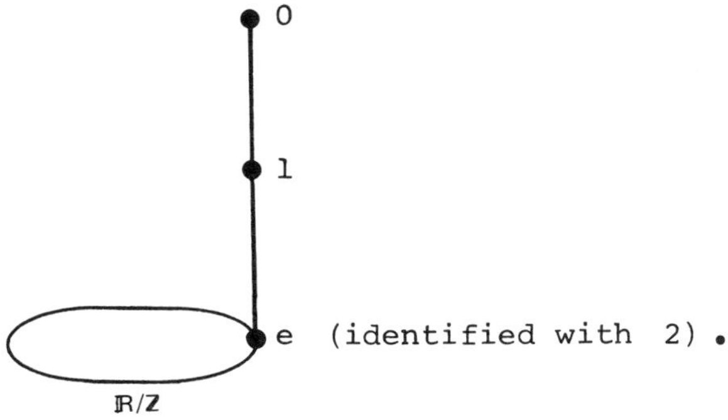

In case $x = \frac{3}{2}$, we have:

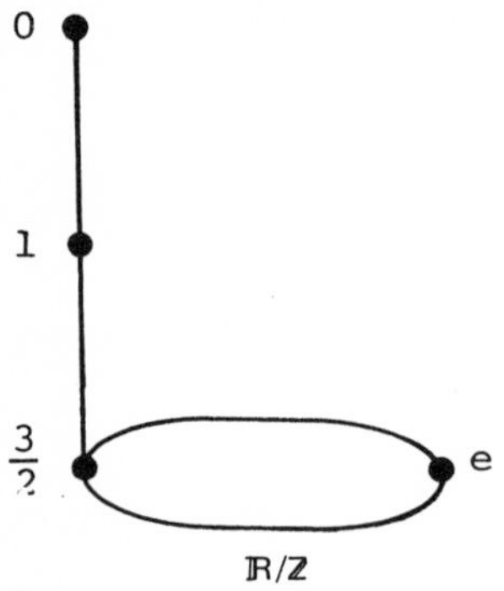

(identified with 2) .

This construction allows one to form numerous "golf club" examples. Observe that for $x \in \mathbb{N}$, the "handle" enters M(S) at the identity of M(S), where $S = \mathbb{H} \underset{\phi,I}{\cup} \mathbb{H}/\mathbb{Z}$, but in no such case is the handle a subsemigroup of S. If J is one of the semigroups I_u, I_n or I_m and T is the subsemigroup of $\mathbb{R}/\mathbb{Z} \times J$ consisting of $\{(g,t) : g = 1$ or $t = 0\}$, then T is a semigroup on the golf club whose handle is a subsemigroup. Clearly, if the handle of a golf club S is a subsemigroup of S, then it must enter M(S) at the identity of M(S).

Our final example of this section illustrates a construction which has been referred to as "replacing the base" (see [Hunter, 1963]).

Let S be a compact semigroup such that M(S) is a group, G a compact, and $\psi : M(S) \to G$ a continuous homomorphism. Define $\phi : S \to G$ by $\phi = \psi \circ \lambda_e$, where e is the identity of M(S). Then ϕ is a continuous homomorphism and the compact semigroup $S \underset{\phi,M(S)}{\cup} G$ has a minimal ideal which is topologically isomorphic to G. In other terms, the minimal ideal of S has been replaced by G.

PROJECTIVE AND INJECTIVE LIMITS

The concepts of projective (or inverse) and injective (or direct) limits of topological semigroups are developed in this section and some results on compact semigroups are obtained. In particular, we will show that each compact totally disconnected semigroup is a projective limit of finite discrete semigroups [Numakura, 1957] and that each compact semigroup is a projective limit of compact metric semigroups [Hofmann and Mostert, 1966]. Our proofs of these and related results were motivated by the above references, by [Eilenberg, 1934], and by the treatment of acts as projective limits of finite and metric acts in [Friedberg, 1973].

A *projective system of [topological] semigroups* is a triple $((D,\leqslant), \{S_\alpha\}_{\alpha\in D}, \{\phi_\alpha^\beta\}_{\alpha\leqslant\beta})$ where:

(a) $(D,\leqslant)$ is a directed set;

(b) $\{S_\alpha\}_{\alpha\in D}$ is a family of [topological] semigroups indexed by D; and

(c) $\{\phi_\alpha^\beta\}_{\alpha\leqslant\beta}$ is a family of functions indexed by $\leqslant$ such that

(i) $\phi_\alpha^\beta : S_\beta \to S_\alpha$ is a [continuous] homomorphism for each $(\alpha,\beta) \in \leqslant$;

(ii) $\phi_\alpha^\alpha = 1_{S_\alpha}$ for each $\alpha \in D$; and

(iii) $\phi_\alpha^\beta \circ \phi_\beta^\gamma = \phi_\alpha^\gamma$ for all $\alpha \leqslant \beta \leqslant \gamma$ in D.

When no confusion is likely, we will denote the projective system $((D,\leqslant), \{S_\alpha\}_{\alpha\in D}, \{\phi_\alpha^\beta\}_{\alpha\leqslant\beta})$ by $\{S_\alpha,\phi_\alpha^\beta\}_{\alpha\in D}$. Each ϕ_α^β is called a *bonding map* and $\{S_\alpha,\phi_\alpha^\beta\}_{\alpha\in D}$ is said to be *strict* if each bonding map is surjective. If $S = \{x \in \Pi\{S_\alpha\}_{\alpha\in D} : \phi_\alpha^\beta(x(\beta)) = x(\alpha)$ for all $\alpha \leqslant \beta$ in $D\}$ is non-empty, then S is called the *projective limit* of $\{S_\alpha,\phi_\alpha^\beta\}_{\alpha\in D}$. If $\{S_\alpha,\phi_\alpha^\beta\}_{\alpha\in D}$ is a strict projective system, then S is called the *strict projective limit* of $\{S_\alpha,\phi_\alpha^\beta\}_{\alpha\in D}$. We write $S = \varprojlim \{S_\alpha,\phi_\alpha^\beta\}_{\alpha\in D}$ or, when no confusion is likely, $S = \varprojlim S_\alpha$.

> *2.20 Theorem*. Let $\{S_\alpha,\phi_\alpha^\beta\}_{\alpha\in D}$ be a projective system of [topological] semigroups such that $S = \lim S$ exists. Then S is a [closed] subsemigroup of $\Pi\{S_\alpha\}_{\alpha\in D}$.

Proof. For fixed $\beta \leqslant \gamma$ in D, let $T_\beta^\gamma = \{x \in \Pi\{S_\alpha\}_{\alpha\in D} : \pi_\beta^\gamma(x(\gamma)) = x(\beta)\}$. Then T_β^γ is readily seen to be a [closed] subsemigroup of $\Pi\{S_\alpha\}_{\alpha\in D}$ and, as $S = \cap \{T_\beta^\gamma : \beta \leqslant \gamma\}$, S is a [closed] subsemigroup of $\Pi\{S_\alpha\}_{\alpha\in D}$. ■

We will show that the projective limit of a projective system of compact semigroups always exists. The next example illustrates that this is not the case for semigroups in general.

Let $S_n = \mathbb{N}$ for each $n \in \mathbb{N}$ and define $\phi_n^m : S_m \to S_n$ for $n \leqslant m$ by $\phi_n^m(k) = 2^{m-n}k$. Then $\{S_n, \phi_n^m\}_{n \in \mathbb{N}}$ is a projective system of (discrete) topological semigroups. Fix $x \in \Pi\{S_n\}_{n\in\mathbb{N}}$ and let $p \in \mathbb{N}$ such that $2^{p-1} > x(1)$. Then $\phi_1^p(x(p)) = 2^{p-1} > x(1)$ and so $x \notin T_1^p = \{y \in \Pi\{S_n\}_{n\in\mathbb{N}} : \phi_1^p(y(p)) = y(1)$. It follows that the projective limit of the projective system $\{S_n, \phi_n^m\}_{n\in\mathbb{N}}$ does not exist.

2.21 Lemma. If $\mathcal{F}$ is a descending family of non-empty subcontinua of a space Y, then $\cap\, \mathcal{F}$ is a non-empty subcontinuum of Y.

Proof. That $\cap\, \mathcal{F}$ is compact and Hausdorff is clear and it is non-empty. Suppose that $\cap\, \mathcal{F}$ is the union of disjoint closed sets P and Q. If U and V are disjoint open sets containing P and Q, respectively, then there is an $F \in \mathcal{F}$ such that $F \subset U \cup V$. As F is connected, $F \subset U$ of $F \subset V$ and it follows that $P = \square$ or $Q = \square$. Hence, $\cap\, \mathcal{F}$ is connected. ■

2.22 Theorem. Let $\{S_\alpha, \phi_\alpha^\beta\}_{\alpha\in D}$ be a projective system of compact [and connected] semigroups. Then $\varprojlim S_\alpha$ is a compact [and connected] semigroup.

Proof. For fixed $\gamma \in D$ let $T^\gamma = \{x \in \Pi\{S_\alpha\}_{\alpha\in D} : \phi_\beta^\gamma(x(\gamma)) = x(\beta)$ for all $\beta \leqslant \gamma$ in D$\}$. For each $x \in \Pi\{S_\alpha\}_{\alpha\in D}$, if f(x) is defined by

$$f(x)(\beta) = \begin{cases} x(\beta) & \text{if } \beta \not\leqslant \gamma \\ \\ \phi_\beta^\gamma(x(\gamma)) & \text{if } \beta \leqslant \gamma \end{cases}$$

for all $\beta \in D$, then $f(x) \in T^\gamma$. Hence, $T^\gamma \neq \square$ and f is a continuous function from $\Pi\{S_\alpha\}_{\alpha\in D}$ into T^γ since

$$\pi_\beta \circ f = \begin{cases} \pi_\beta & \text{if } \beta \not\leqslant \gamma \\ \phi_\beta^\gamma \circ \pi_\gamma & \text{if } \beta \leqslant \gamma. \end{cases}$$

It is obvious that f retracts $\Pi\{S_\alpha\}_{\alpha\in D}$ onto T^γ and so T^γ is a compact [and connected] subspace of $\Pi\{S_\alpha\}_{\alpha\in D}$. Moreover, if $\gamma \leqslant \delta$ in D, then $T^\delta \subset T^\gamma$ and so $\{T^\gamma\}_{\gamma\in D}$ is a descending family of compact [and connected] sets. Therefore, $\varprojlim S_\alpha = \cap\{T^\gamma : \gamma \in D\}$ is a non-empty compact [and connected (2.21)] subspace of $\Pi\{S_\alpha\}_{\alpha\in D}$. The conclusions now follow from 2.20. ∎

Observe that if $\{S_\alpha, \phi_\alpha^\beta\}_{\alpha\in D}$ is a projective system of abelian semigroups, then $\varprojlim S_\alpha$ is abelian if it exists.

2.23 Theorem. Let $\{S_\alpha, \phi_\alpha^\beta\}_{\alpha\in D}$ be a strict projective system of compact semigroups and let $S = \varprojlim S_\alpha$. Then $(\pi_\alpha|S) : S \to S_\alpha$ is surjective for each $\alpha \in D$.

Proof. Fix $x_\delta \in S_\delta$ and for $\gamma \geqslant \delta$ in D, let $B^\gamma = \{x \in \Pi\{S_\alpha\}_{\alpha\in D} : \phi_\alpha^\gamma(x(\gamma)) = x(\beta)$ for all $\beta \leqslant \gamma$ and $x(\delta) = x_\delta\}$. As ϕ_δ^γ is surjective, let x_γ be chosen in S_γ such that $\phi_\delta^\gamma(x_\gamma) = x_\delta$. For $x \in \Pi\{S_\alpha\}_{\alpha\in D}$, define f(x) by

$$f(x)(\beta) = \begin{cases} x(\beta) & \text{if } \beta \not\leqslant \gamma \\ \phi_\beta^\gamma(x_\gamma) & \text{if } \beta \leqslant \gamma. \end{cases}$$

Then, $f(x) \in B^\gamma$ and f is continuous since

$$\pi_\beta \circ f = \begin{cases} \pi_\beta & \text{if } \beta \not\leqslant \gamma \\ \phi_\beta^\gamma \circ c \circ \pi_\gamma & \text{if } \beta \leqslant \gamma \end{cases}$$

for all $\beta \in D$ (where c is the constant map from S_γ to $\{x_\gamma\}$). As in the proof of 2.2, $\{B^\gamma\}_{\gamma\geqslant\delta}$ is a descending family of non-empty compact subspaces of $\Pi\{S_\alpha\}_{\alpha\in D}$. If $x \in \cap \{B^\gamma : \gamma \geqslant \delta \text{ in } D\}$, then $x \in \varprojlim S_\alpha$ and $\pi_\delta(x) = x_\delta$. ■

We now turn our attention to the following question: Given a class $\mathcal{C}$ of semigroups, what semigroups are isomorphic to [strict] projective limits of members of $\mathcal{C}$?

Recall that if S is a semigroup, $\mathcal{R} = \{R_\alpha\}_{\alpha\in D}$ is a descending family of congruences on S, and $R = \cap \mathcal{R}$, then π, $\{\pi_\alpha\}_{\alpha\in D}$, $\{\phi_\alpha\}_{\alpha\in D}$, $\{\phi_\beta^\gamma\}_{\beta\leqslant\gamma}$, and ψ are defined by the diagrams:

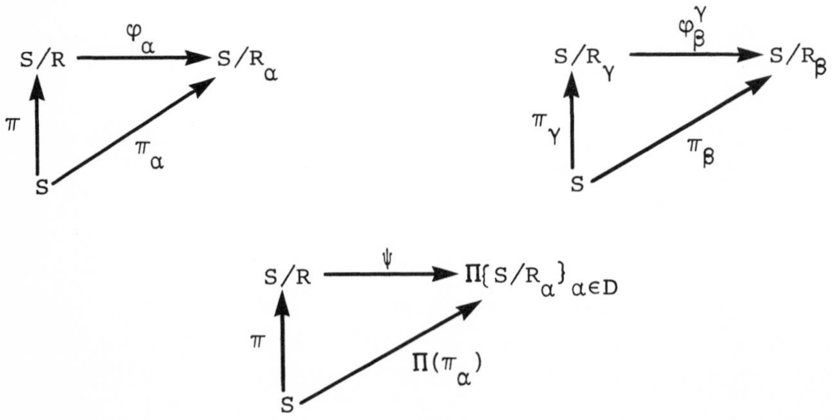

The non-empty family $\mathcal{R}$ of subsets of $S \times S$ *separates points of* S if $\cap \mathcal{R} \subset \Delta$.

<u>*2.24 Theorem*</u>. Let $\mathcal{R} = \{R_\alpha\}_{\alpha\in D}$ be a descending family of closed congruences on the compact semigroup S. Then

(a) $\{S/R_\alpha, \phi_\alpha^\beta\}_{\alpha\in D}$ is a strict projective system of compact semigroups;

(b) ψ is a topological isomorphism from S/R onto $\varprojlim S/R_\alpha$; and

(c) If $\mathcal{R}$ separates points of S, then S is topologically isomorphic to $\varprojlim S/R_\alpha$ under the map $\psi\pi$.

Proof. (a) follows immediately from the definitions. For (b), observe from 2.3 that ψ is a topological isomorphism from S/R into

$\Pi\{S/R_\alpha\}_{\alpha\in D}$. If $\pi(a) \in S/R$, then $\phi_\beta^\gamma((\psi\pi(a))(\gamma)) = \phi_\beta^\gamma(\pi_\gamma(a)) = \pi_\beta(a) = (\psi\pi(a)(\beta)$ and so $\phi\pi(a) \in \underset{\leftarrow}{\lim} S/R_\alpha$. For the other containment, fix $f \in \underset{\leftarrow}{\lim} S/R_\alpha$ and for $\alpha \in D$, let $L_\alpha = \pi_\alpha^{-1}(f(\alpha))$. Then $\{L_\alpha\}_{\alpha\in D}$ is a descending family of compact subsets of S. If α is chosen in $\cap \{L_\alpha : \alpha \in D\}$, then $\pi(a) \in S/R$ and $\pi_\alpha(a) = f(\alpha)$ for all $\alpha \in D$. Thus, $\psi\pi(a) = f$ and ψ is onto $\underset{\leftarrow}{\lim} S/R_\alpha$. Finally, (c) is clear, as π is a topological isomorphism if $\cap \mathcal{R} = \Delta$. ■

2.25 Lemma. Let R be an open and closed congruence on the compact semigroup A. Then S/R is a finite discrete semigroup.

Proof. If $\pi : S \to S/R$ is the natural map, then for each $\pi(a) \in S/R$, $\pi^{-1}\pi(a) = \{x \in S : (x,a) \in R\}$ is open since R is open. Hence, $\pi(a)$ is open and S/R is discrete. However, S/R is also compact and so S/R is finite. ■

The following result appears in [Numakura, 1957].

2.26 Theorem. Let S be a compact totally disconnected semigroup. Then S is (topologically isomorphic to) a strict projective limit of finite discrete semigroups.

Proof. Let $\mathcal{R} = \{R : R$ is an open and closed congruence on $S\}$. Clearly $\mathcal{R}$ is a descending family. In view of 2.24 and 2.25, it suffices to show that $\mathcal{R}$ separates points of S. To that end, let x and y be distinct points in S. According to 1.58 there is an open and closed set K such that $x \in K \subset S\backslash\{y\}$. If $\mathcal{V} = (K \times K) \cup ((S\backslash K) \times (S\backslash K))$, then $\mathcal{V}$ is open and closed and so $\mathcal{U} = \cup \{\mathcal{W} : \mathcal{W}$ is a Δ-ideal and $\mathcal{W} \subset \mathcal{V}\}$ is an open and closed Δ-ideal. Obviously $\Delta \subset \mathcal{U}$. Now, $\mathcal{U}^{-1}$ is a Δ-ideal, and since $\mathcal{V}$ is symmetric, $\mathcal{U}^{-1} \subset \mathcal{V}$ and so $\mathcal{U}^{-1} \subset \mathcal{U}$. Thus $\mathcal{U}^{-1} = \mathcal{U}$ and $\mathcal{U}$ is symmetric. Similarly, $\mathrm{Tr}(\mathcal{U})$ is a Δ-ideal and since $\mathcal{V}$ is transitive, $\mathrm{Tr}(\mathcal{U}) = \mathcal{U}$ and so $\mathcal{U}$ is transitive. Finally, $\mathcal{U}$ is a compatible relation as it is a Δ-ideal. Therefore, $\mathcal{U} \in \mathcal{R}$, $(x,y) \in (S \times S)\backslash\mathcal{U}$, and the proof is complete. ■

A metric d on a semigroup S is said to be *subinvariant* if for each a, x, y $\in$ S we have $d(ax,ay) \leqslant d(x,y)$ and $d(xa,ya) \leqslant d(x,y)$.

A topological semigroup S is said to be a *metric semigroup* if there exists a subinvariant metric d on S which determines the topology of S.

2.27 Theorem. If the space of a compact semigroup S is metrizable, then S is a metric semigroup.

Proof. Let ρ be a bounded metric on S which determines the topology of S and define $d(x,y) = \sup \{\rho(axb,ayb) : a, b \in S^{-1}\}$ for $x, y \in S$. Then it is straightforward to show that d is a subinvariant metric on S and it is clear that the identity function 1_S is continuous from (S,ρ) into (S,d). By the compactness of (S,ρ), 1_S is a homeomorphism and so d determines the topology of S. ■

2.28 Lemma. Let S be a compact semigroup and let f be a continuous function from S into [0,1]. Then $R_f = \{(x,y) \in S \times S : f(axb) = f(ayb)$ for all $a, b \in S^1\}$ is a closed congruence on S and S/R_F is a metric semigroup.

Proof. First we observe that $R_f = \cup \{W : W$ is a Δ-ideal and $W \subset (f \times f)^{-1}(\Delta)\}$. Now, using 1.38 and 1.39 as in the proof of 2.26, we obtain that R_f is a closed congruence on S. If $\pi : S \to S/R_f$ is the natural map, define d on $(S/R_f) \times (S/R_f$ by $d(\pi(x),\pi(y)) = \sup \{|f(axb) - f(ayb)| : a, b \in S^1\}$. Again, it is straightforward to show that d is a well-defined metric on S/R_f (in fact, d is subinvariant). If $\pi(x) \in S/R_f$, $\varepsilon > 0$, and $A = \{\pi(y) : d(\pi(x),\pi(y)) < \varepsilon\}$ is the associated basis metric open set, then it is not difficult to show that $B = \pi^{-1}(A) = \{y \in S : |f(axb) - f(ayb)| < \varepsilon$ for all $a, b \in S^1\}$. This set is open in S as follows: Define $g : S^1 \times S \times S^1 \to [0,1]$ by $g(a,y,b) = |f(axb) - b(ayb)|$. Then g is continuous and $y \in B$ if and only if $g[S^1 \times \{y\} \times S^1] \subset [0,\varepsilon)$. Now, by 1.1, for each $y \in B$ there exists an open set U such that $y \in U$ and $g[S^1 \times U \times S^1] \subset [0,\varepsilon)$. Hence, $y \in U \subset B$, B is open in S, and A is open in the quotient topology. It follows that d determines the topology of S/R_f. ■

The next result appears in [Hofmann and Mostert, 1966].

2.29 Theorem. Let S be a compact semigroup. Then S is (topologically isomorphic to) a strict projective limit of compact metric semigroups.

Proof. Let $\mathcal{R}$ denote the family of all closed congruences R on S such that S/R is a compact metric semigroup. If $R_1, R_2 \in \mathcal{R}$ then $R_1 \cap R_2$ is a closed congruence on S. By 1.7, $S/(R_1 \cap R_2)$ can be embedded into the compact metrizable semigroup $(S/R_1) \times (S/R_2)$. Hence, $S/(R_1 \cap R_2)$ is metrizable and, according to 2.27, $S/(R_1 \cap R_2)$ is a metric semigroup. That is, $R_1 \cap R_2 \in \mathcal{R}$ and R is a descending family. If x and y are distinct points of S, let $f : S \to [0,1]$ be a map such that $f(x) = 0$ and $f(y) = 1$. Then $R_f \in \mathcal{R}$ (2.28) and $(x,y) \in (S \times S)\backslash R_f$. Therefore, $\mathcal{R}$ separates points of S. The conclusion of the theorem now follows from 2.24. ∎

Observe that the results of 2.20, 2.22, 2.23, 2.25, 2.26, and 2.29 remain valid if we replace "semigroup" by "group" throughout (e.g. the projective limit of compact groups is again a compact group). It should also be observed that a subinvariant metric d on a group G is in fact invariant; that is, $d(ab,ac) = d(ba,ca) = d(b,c)$ for all a, b, $c \in G$. To see this, suppose d is a subinvariant metric on a group G. Then for a, b, $c \in G$ we have

$$d(ab,ac) \leqslant d(b,c) = d(a^{-1}ab,a^{-1}ac) \leqslant d(ab,ac)$$

so that $d(ab,ac) = d(b,c)$. Similarly, $d(ba,ca) = d(b,c)$.

Before considering some examples of projective systems, we discuss a technique for constructing some sequential projective systems.

Suppose that $\{S_n\}_{n\in\mathbb{N}}$ is a sequence of compact semigroups and $\{\phi_n\}_{n\in\mathbb{N}}$ is a sequence of continuous surmorphisms, where $\phi_n : S_{n+1} \to S_n$ for each $n \in \mathbb{N}$. Define $\phi_n^m : S_m \to S_n$ for $n < m$ in $\mathbb{N}$ by $\phi_n^m = \phi_n^{n+1} \circ \cdots \circ \phi_{m-1}^m$ where $\phi_k^{k+1} = \phi_k$ and $\phi_n^n = 1_{S_n}$ for each $n \in \mathbb{N}$. Then $\{S_n,\phi_n^m\}_{n\in\mathbb{N}}$ is a strict projective system of compact semigroups and the projective limit is denoted $\varprojlim \{S_n,\phi_n\}_{n\in\mathbb{N}}$.

Let $G_n = \mathbb{R}/\mathbb{Z}$ for each $n \in \mathbb{N}$ and let $a = \{a_n\}_{n\in\mathbb{N}}$ be a sequence of positive integers. Define $\phi_n : G_{n+1} \to G_n$ by $\phi_n(z) = z^{a_n}$

for each $n \in \mathbb{N}$ and each $z \in G_{n+1}$. The compact abelian group $\varprojlim \{G_n, \phi_n\}_{n\in\mathbb{N}}$ is called the *a-adic solenoid* and is denoted by Σ_a.

Let $S_j = I_n$ (the nil interval) for each $j \in \mathbb{N}$ and define $\phi_j : S_{j+1} \to S_j$ by $\phi_j(x) = x^j$ for each $x \in S_{j+1}$. Then $I_u \overset{T}{\approx} \varprojlim \{S_j, \phi_j\}_{j\in\mathbb{N}}$ [Brown, 1960].

Finally, we turn our attention to injective systems and injective limits. At this point we will give the definition of an injective system of topological semigroups.

An injective system of [*topological*] *semigroups* is a triple $((D,\leqslant), \{S_\alpha\}_{\alpha\in D}, \{\phi_\alpha^\beta\}_{\alpha\leqslant\beta})$ where:

(a) $(D,\leqslant)$ is a directed set;

(b) $\{S_\alpha\}_{\alpha\in D}$ is a family of [topological] semigroups indexed by D; and

(c) $\{\phi_\alpha^\beta\}_{\alpha\leqslant\beta}$ is a family of functions indexed by $\leqslant$ such that:

(i) $\phi_\alpha^\beta : S_\alpha \to S_\beta$ is a [continuous] homomorphism for each $(\alpha,\beta) \in \leqslant$;

(ii) $\phi_\alpha^\alpha = 1_{S_\alpha}$ for each $\alpha \in D$; and

(iii) $\phi_\beta^\gamma \circ \phi_\alpha^\beta = \phi_\alpha^\gamma$ for all $\alpha \leqslant \beta \leqslant \gamma$ in D.

When no confusion is likely, we will denote the injective system $((D,\leqslant), \{S_\alpha\}_{\alpha\in D}, \{\phi_\alpha^\beta\}_{\alpha\leqslant\beta}$ by $\{S_\alpha, \phi_\alpha^\beta\}_{\alpha\in D}$.

<u>*2.30 Lemma*</u>. Let $\{S_\alpha, \phi_\alpha^\beta\}_{\alpha\in D}$ be an injective system of semigroups and let $A = \cup \{\{\alpha\} \times S_\alpha : \alpha \in D\}$. Then $R = \{((\alpha,x),(\beta,y)) \in A \times A : \phi_\alpha^\gamma(x) = \phi_\beta^\gamma(y)$ for some $\gamma \geqslant \alpha, \beta$ in $D\}$ is an equivalence on A.

Proof. Clearly R is reflexive and symmetric. Suppose now that $((\alpha,x),(\beta,y)) \in R$ and $((\beta,y),(\gamma,z)) \in R$. Fix $\delta \geqslant \alpha, \beta$ and $\theta \geqslant \beta, \gamma$ in D such that $\phi_\alpha^\delta(x) = \phi_\beta^\delta(y)$ and $\phi_\beta^\theta(y) = \phi_\gamma^\theta(z)$. Let ψ be chosen in D so that $\psi \geqslant \delta, \theta$. Then

$$\phi_\alpha^\psi(x) = \phi_\delta^\psi \circ \phi_\alpha^\delta(x) = \phi_\delta^\psi \circ \phi_\beta^\delta(y) = \phi_\beta^\psi(y) = \phi_\theta^\psi \circ \phi_\beta^\theta(y)$$

$$= \phi_\theta^\psi \circ \phi_\delta^\theta(z) = \phi_\gamma^\psi(z)$$

and so $((\alpha,x),(\gamma,z)) \in R$. Thus R is transitive and the desired result is obtained. ∎

2.31 Theorem. Let $\{S_\alpha, \phi_\alpha^\beta\}_{\alpha\in D}$, A, and R be as in 2.30 and denote A/R by S. Let $\pi : A \to S$ be the natural map and define a subset m of $(S \times S) \times S$ by letting $((a,b),c) \in m$ if and only if there exist (α,x), (β,y), and (γ,z) in A such that $a = \pi(\alpha,x)$, $b = \pi(\beta,y)$, $c = \pi(\gamma,z)$, $\gamma \geqslant \alpha, \beta$, and $\phi_\alpha^\gamma(x)\phi_\beta^\gamma(y) = z$. Then:

(a) m is a function from $S \times S$ into S;

(b) (S,m) is a semigroup; and

(c) If S_α is a group for each α, then (S,m) is a group.

Proof. For $(a,b) \in S \times S$, fix (α,x) and (β,y) in A such that $\pi(\alpha,x) = a$ and $\pi(\beta,y) = b$. If γ is chosen in D so that $\gamma \geqslant \alpha, \beta$, then $((a,b),\phi_\alpha^\gamma(x)\phi_\beta^\gamma(y)) \in m$ and so $\pi_1(m) = S \times S$. Now suppose $((a,b)c) \in m$ and $((a,b),d) \in m$. Fix (α,x), (β,y), (γ,z), (ρ,r), (σ,s), $(\tau,t) \in A$ such that $a = \pi(\alpha,x) = \pi(\rho,r)$, $b = \pi(\beta,y) = \pi(\sigma,s)$, $c = \pi(\gamma,z)$, $d = \pi(\tau,t)$, $\gamma \geqslant \alpha, \beta$, $\tau \geqslant \rho, \sigma$, $\phi_\alpha^\gamma(x)\phi_\beta^\gamma(y) = z$, and $\phi_\rho^\tau(r)\phi_\sigma^\tau(s) = t$. Then there exist $\lambda, \mu \in D$ such that $\gamma \geqslant \alpha, \rho$, $\mu \geqslant \beta, \sigma$, $\phi_\alpha^\lambda(x) = \phi_\rho^\lambda(r)$, and $\phi_\beta^\mu(y) = \phi_\sigma^\mu(s)$. Let δ be chosen in D such that $\delta \geqslant \lambda, \mu, \gamma, \tau$. Then

$$\begin{aligned}\phi_\gamma^\delta(z) &= \phi_\gamma^\delta(\phi_\alpha^\gamma(x)\phi_\beta^\gamma(y)) = \phi_\gamma^\delta\phi_\alpha^\gamma(x)\phi_\gamma^\delta\phi_\beta^\gamma(y) = \phi_\alpha^\delta(x)\phi_\beta^\delta(y)\\ &= \phi_\lambda^\delta\phi_\alpha^\lambda(x)\phi_\mu^\delta\phi_\beta^\mu(y) = \phi_\lambda^\delta\phi_\rho^\lambda(r)\phi_\mu^\delta\phi_\sigma^\mu(s) = \phi_\rho^\delta(r)\phi_\sigma^\delta(s)\\ &= \phi_\tau^\delta\phi_\rho^\tau(r)\phi_\tau^\delta\phi_\sigma^\tau(s) = \phi_\tau^\delta(\phi_\rho^\tau(r)\phi_\sigma^\tau(s)) = \phi_\tau^\delta(t).\end{aligned}$$

Hence, $((\gamma,z),(\tau,t)) \in R$ and so $c = \pi(\gamma,z) = \pi(\tau,t) = d$. This completes the proof of (a).

For (b), denote multiplication by juxtaposition and observe that if $a = \pi(\alpha,x)$ and $b = \pi(\beta,y)$, then $ab = \pi(\delta,\phi_\alpha^\delta(x)\phi_\beta^\delta(y))$ for each $\delta \geqslant \alpha, \beta$ in D. Now, fix a, b, $c \in S$ and choose (α,x), (β,y), and (γ,z) in A such that $a = \pi(\alpha,x)$, $b = \pi(\beta,y)$, and $c = \pi(\gamma,z)$. Then, for a fixed $\delta \geqslant \alpha, \beta, \gamma$ we have

$$(ab)c = (\pi(\alpha,x)\pi(\beta,y))\pi(\gamma,z) = \pi(\delta,\phi_\alpha^\delta(x)\phi_\beta^\delta(y))\pi(\gamma,z)$$
$$= \pi(\delta,(\phi_\alpha^\delta(x)\phi_\beta^\delta(y))\phi_\gamma^\delta(z)) = \pi(\delta,\phi_\alpha^\delta(x)(\phi_\beta^\delta(y)\phi_\gamma^\delta(z)))$$
$$= \pi(\alpha,x)\pi(\delta,\phi_\beta^\delta(y)\phi_\gamma^\delta(z)) = \pi(\alpha,x)(\pi(\beta,y)\pi(\gamma,z)) = a(bc).$$

Finally, suppose that each S_α is a group and denote the identity of S_α by e_α for all $\alpha \in D$. If $\alpha, \beta \in D$ and $\delta \geqslant \alpha, \beta$, then

$$\pi(\alpha,e_\alpha) = \pi(\delta,\phi_\alpha^\delta(e_\alpha)) = \pi(\delta,e_\delta) = \pi(\delta,\phi_\beta^\delta(e_\beta)) = \pi(\beta,e_\beta).$$

Let e denote the R-class of A containing all elements of the form (α,e_α). Now, if $\alpha \in S$ and $a = \pi(\alpha,x)$, then $ea = \pi(\alpha,e_\alpha)\pi(\alpha,x) = \pi(\alpha,e_\alpha x) = \pi(\alpha,x) = a$. Thus, e is a left identity for S. Also, if $a \in S$ and $a = \pi(\alpha,x)$, then letting x^{-1} denote the inverse of x in S_α we have

$$\pi(\alpha,x^{-1})a = \pi(\alpha,x^{-1})\pi(\alpha,x) = \pi(\alpha,x^{-1}x) = \pi(\alpha,e_\alpha) = e.$$

Therefore, each element of S has a left inverse relative to e. It follows that (S,m) is a group [van der Waerden, 1949]. ∎

The semigroup (S,m) of 2.31 is called the injective limit of the injective system $\{S_\alpha,\phi_\alpha^\beta\}_{\alpha\in D}$. This semigroup is denoted by $\underleftarrow{\lim}\,\{S_\alpha,\phi_\alpha^\beta\}_{\alpha\in D}$, or when no confusion seems likely, by $\underleftarrow{\lim}\, S_\alpha$.

Let $\{S_m\}_{m\in\mathbb{N}}$ be an increasing sequence of semigroups (i.e., $S_m \subset S_{m+1}$ for each $m \in \mathbb{N}$) and let $\phi_m^n : S_m \to S_n$ be the inclusion of S_m in S_n for $m \leqslant n$. Then $\{S_m,\phi_m^n\}_{m\in\mathbb{N}}$ is an injective system of semigroups. If $S = \cup\,\{S_m : m \in \mathbb{N}\}$ with the obvious multiplication, then $S \approx \underleftarrow{\lim}\, S_m$.

SEMIGROUPS OF HOMOMORPHISMS

The culmination of this section is a brief survey of Pontryagin duality theory for locally compact abelian groups. The approach taken here is to place the compact-open topology on the semigroup of continuous homomorphisms of a locally compact semigroup into an abelian topological semigroup (with pointwise multiplication), thus obtaining an abelian topological semigroup. The more elementary parts of the development are treated in detail and the deeper results are left to other sources, e.g., the duality theorem itself. As it happens, duality theory is basic to the study of Bohr compactifications of certain semigroups and to the study of irreducible semigroups.

The proof of the following is straightforward and is omitted.

2.32 Lemma. Let S be a semigroup and let T be an abelian semigroup. If f and g are homomorphisms from S into T and fg is defined by $(fg)(x) = f(x)g(x)$ for all $x \in S$, then fg is also a homomorphism from S into T. Hence, if the set of homomorphisms from S into T is nonempty, then it is an abelian semigroup under this multiplication.

For topological semigroups S and T with T abelian, we use Hom (S,T) to denote the set of all continuous homomorphisms from S into T with pointwise multiplication (the multiplication of 2.32) and the compact-open topology. If Hom $(S,T) \neq \square$, then according to 2.32, Hom (S,T) is an abelian semigroup.

If S and T are topological semigroups with T abelian and $E(T) \neq \square$, then Hom $(S,T) \neq \square$ as $e \in E(T)$ and $f : S \to T$ the constant map with image $\{e\}$ imply $f \in$ Hom (S,T). In particular, if T is compact or T is a group, then Hom $(S,T) \neq \square$.

2.33 Theorem. Let S be a locally compact semigroup and let T be an abelian topological semigroup such that Hom $(S,T) \neq \square$. Then Hom (S,T) is an abelian topological semigroup.

Proof. Since Hom (S,T) is a subspace of the Hausdorff space C(S,T), we have that Hom (S,T) is a Hausdorff space (see the proof of 2.6). In view of 2.32, it remains to be shown that multiplication on Hom (S,T) is continuous. To this end suppose that f, g $\in$ Hom (S,T) and fg $\in N(K,W)$, where K is a compact subspace of S and W is an open subset of T. Then $(fg)(K) \subset W$ and so $f(x)g(x) \in W$ for each $x \in K$. Using continuity of multiplication in T, for each $x \in K$, there exists open sets U_x and V_x in T such that $f(x) \in U_x$, $g(x) \in V_x$, and $U_x V_x \subset W$. In view of the local compactness of S and the continuity of f and g, for each $x \in K$, there exist open sets A_x and C_x and a compact set B_x such that $x \in A_x \subset B_x \subset C_x$, $f(C_x) \subset U_x$, and $g(C_x) \subset V_x$. Since K is compact, there exists a finite cover $A_1,\ldots,A_n$ of the A_x's and corresponding $B_1,\ldots,B_n$ of B_x's. Let $U_1,\ldots,U_n$ and $V_1,\ldots,V_n$ be the corresponding collections of U_x's and V_x's, respectively. Then $B_1,\ldots,B_n$ is a cover of K by compact sets, and for $1 \leqslant j \leqslant n$, we have $f(B_j) \subset U_j$, $g(B_j) \subset V_j$, and $U_j V_j \subset W$. Observe that $f \in \bigcap_{j=1}^{n} N(B_j,U_j)$ and $g \in \bigcap_{j=1}^{n} N(B_j,V_j)$. Suppose $h \in \bigcap_{j=1}^{n} N(B_j,U_j)$ and $k \in \bigcap_{j=1}^{n} N(B_j,V_j)$ and fix $x \in K$. Then $x \in B_i$ for some $1 \leqslant i \leqslant n$ and so $(hk)(x) = h(x)k(x) \in U_i V_i \subset W$. It follows that $hk \in N(K,W)$, and hence

$$\left(\bigcap_{j=1}^{n} N(B_j,U_j)\right)\left(\bigcap_{j=1}^{n} N(B_j,V_j)\right) \subset N(K,W). \quad \blacksquare$$

2.34 Theorem. Let S be a locally compact semigroup and let T be an abelian topological group. Then Hom (S,T) is an abelian topological group.

Proof. In view of 2.32 and 2.33, we need only establish that Hom (S,T) is a group and that inversion is continuous. Let 1 denote the identity of T and $\phi : T \to T$ inversion, i.e., $\phi(x) = x^{-1}$ for each $x \in T$. Define $e : S \to T$ by $e(x) = 1$ for all $x \in S$. Then $ef = fe = f$ for all $f \in$ Hom (S,T) and $e \in$ Hom (S,T) so that e is an identity for Hom (S,T). If $f \in$ Hom (S,T) define $\tilde{\tilde{f}} : S \to T$ by $\tilde{f} = \phi \circ f$. Then $\tilde{f} \in$ Hom (S,T) for each $f \in$ Hom (S,T) and the function ρ : Hom (S,T) $\to$ Hom (S,T) defined by $\rho(f) = \tilde{f}$ is inversion. Thus, Hom (S,T) is a group. To show that ρ is continuous, let K be a compact subset of S, W an open subset of T, $f \in$ Hom (S,T), and $\rho(f) \in N(K,W)$. Then $(\phi \circ f)(K) = \tilde{f}(K) = \rho(f)(K) \subset W$ and hence $f \in N(K,\phi^{-1}(W))$. Then $g(K) \subset \phi^{-1}(W)$ and $(g)(K) = \tilde{\tilde{g}}(K) = \phi \circ g)(K) \subset W$. It follows that $\rho(g) \in N(K,W)$, $\rho(N(K,\phi^{-1}(W)) \subset N(K,W)$, and ρ is continuous. $\blacksquare$

If S is a locally compact semigroup, then a continuous homomorphism from S into $\mathbb{D}$ is called a *semicharacter of* S and Hom $(S,\mathbb{D})$ is called the *semicharacter semigroup of* S. A continuous homomorphism from S into $\mathbb{R}/\mathbb{Z}$ is called a *character* of S and Hom $(S,\mathbb{R}/\mathbb{Z})$ is called the *character group of* S. If, in addition, S is an abelian group, then Hom $(S,\mathbb{R}/\mathbb{Z})$ is also denoted by $S^{\wedge}$ and is called the *dual group of* S.

If T is a fixed abelian topological semigroup, then not only does one naturally associate to each locally compact semigroup S the topological semigroup Hom (S,T), but to each continuous homomorphism $f : S_1 \to S_2$ of locally compact semigroups S_1 and S_2 one naturally associates a continuous homomorphism $\overline{f}_T$: Hom $(S_2,T) \to$ Hom (S_1,T). The definition and basic properties of these associated homomorphisms are summarized next.

If T is an abelian topological semigroup, S_1 and S_2 are locally compact semigroups, and $f : S_1 \to S_2$ is a continuous homomorphism, then we use $\overline{f}_T$ to denote the function from Hom (S_2,T) to T^{S_1} defined by $\overline{f}_T(\phi) = \phi \circ f$ for each $\phi \in$ Hom (S_2,T). When T is fixed throughout a discussion and no confusion is likely, we use simply $\overline{f}$ for $\overline{f}_T$.

2.35 Theorem. Let T be an abelian topological semigroup and let S, S_1, S_2, and S_3 be locally compact semigroups. Let $f : S_1 \to S_2$ and $g : S_2 \to S_3$ be continuous homomorphisms and let 1_S denote the identity homomorphism on S. Then:

(a) $\overline{f}$ is a continuous homomorphism from Hom (S_2,T) into Hom (S_1,T);

(b) $\overline{1_S}$ is the identity homomorphism on Hom (S,T); and

(c) $\overline{g \circ f} = \overline{f} \circ \overline{g}$.

Proof. For (b), we have $\overline{1_S}(\phi) = \phi \circ 1_S = \phi$ for all $\phi \in$ Hom (S,T).

For (c), we have $\overline{g \circ f}(\phi) = \phi \circ (g \circ f) = (\phi \circ g) \circ f = \overline{f}(\phi \circ g) = \overline{f}(\overline{g}(\phi)) = (\overline{f} \circ \overline{g})(\phi)$ for all $\phi \in$ Hom (S_3,T).

To prove (a) first observe that $\overline{f}(\phi) = \phi \circ f$ is a continuous homomorphism from S_1 into T for each $\phi \in \text{Hom}(S_2,T)$, and hence $\overline{f}$: $\text{Hom}(S_2,T) \to \text{Hom}(S_1,T)$. Now, for ϕ and ψ in $\text{Hom}(S_2,T)$ and $x \in S_1$ we have $\overline{f}(\phi\psi)(x) = (\phi\psi)(f(x)) = \phi(f(x))\psi(f(x)) = \overline{f}(\phi)(x)\overline{f}(\psi)(x) = (\overline{f}(\phi)\overline{f}(\psi))(x)$, and it follows that $\overline{f}$ is a homomorphism. Finally, we show that $\overline{f}$ is continuous. Suppose that $\phi \in \text{Hom}(S_2,T)$, K is a compact subset of S_1, W is an open subset of T, and $\overline{f}(\phi) \in N(K,W)$. Then $f(K)$ is a compact subset of S_2, $\phi \in N(f(K),W)$, and it is readily verified that $\overline{f}(N(f(K),W) \subset N(K,W)$. ∎

If G and H are locally compact abelian groups and T is the circle group $\mathbb{R}/\mathbb{Z}$, then for each continuous homomorphism $f : G \to H$, the continuous homomorphism $\overline{f} : H^\wedge \to G^\wedge$ is denoted by $f^\wedge$ and is called the *dual* (or *adjoint*) homomorphism of f.

Next we turn to the development of additional properties of certain semigroups of homomorphisms.

2.36 Theorem. Let S be a discrete semigroup and let T be a compact abelian semigroup. Then $\text{Hom}(S,T)$ is a compact abelian semigroup.

Proof. Suppose that K is a compact subset of S and W is an open subset of T. Since S is discrete we have that K is finite. Let $K = \{k_1,\ldots,k_n\}$ for $n \in \mathbb{N}$. Then $N(K,W) = \bigcap_{i=1}^{n} N(\{k_i\},W)$ and hence $\{N(\{p\},W) : p \in S \text{ and } W \text{ open in } T\}$ is a subbase for the topology of $\text{Hom}(S,T)$. Let $T_p = T$ for each $p \in S$ and consider the compact space $X = \Pi\{T_p\}_{p\in S}$. Define $F : \text{Hom}(S,T) \to X$ by $F(\phi)(x) = \phi(x)$ for all $\phi \in \text{Hom}(S,T)$ and $x \in S$. Then it is clear that F is injective. Now, for the subbasic open set $N(\{q\},W)$ of $\text{Hom}(S,T)$ we have $F(N(\{q\},W)) = F(\{\phi \in \text{Hom}(S,T) : \phi(q) \in W\}) = \{F(\phi) : \phi \in \text{Hom}(S,T) \text{ and } \phi(q) \in W\} = F(\text{Hom}(S,T)) \cap \pi_q^{-1}(W)$; and it follows that F maps $\text{Hom}(S,T)$ homeomorphically into X. Notice that, in view of the discreteness of S, $F(\text{Hom}(S,T))$ is precisely the set of elements of X which are homomorphisms of S into T. We now show that $F(\text{Hom}(S,T))$ is closed in S. Fix $g \in X \backslash F(\text{Hom}(X,T))$. Then $g(xy) \neq g(x)g(y)$ for some $x, y \in S$. Thus there exist open sets W_1, W_2, and W_3 containing g(xy), g(x), and g(y), respectively, such that $W_1 \cap W_2W_3 = \Box$. It follows that $g \in \pi_{xy}^{-1}(W_1) \cap \pi_x^{-1}(W_2) \cap \pi_y^{-1}(W_3) \subset X\backslash F(\text{Hom}\ (S,T))$. Therefore, Hom (S,T) is homeomorphic to a closed subspace of the compact space X and Hom (S,T) is compact. ■

A topological group G is said to have *no small subgroups* if there exists an open set U in G containing the identity of G such that U contains no non-degenerate subgroup of G.

2.37 Theorem. Let S be a compact semigroup and let T be an abelian topological group with no small subgroups. Then Hom (S,T) is a discrete abelian group.

Proof. According to 2.34, we have that Hom (S,T) is an abelian topological group. Let 1 denote the identity of T, e the (constantly 1) identity of Hom (S,T), and U an open subset of T containing 1 and no other subgroup of T. We will show that N(S,U) = {e}. It is clear that e ∈ N(S,U). Suppose that f ∈ N(S,U). Then f(S) is a compact subsemigroup of T and so f(S) is a subgroup of T. Since f(S) ⊂ U, we have that f(S) = {1}, or equivalently f = e. It follows that N(S,U) = {e}. Finally, Hom (S,T) is homogeneous and so Hom (S,T) is discrete. ■

It is well known that a locally compact group is a Lie group if and only if it has no small subgroups (see [Montgomery and Zippin, 1955]). Thus, by 2.37, if S is a compact semigroup and T is an abelian Lie group, then Hom (S,T) is discrete. One can see directly that the group of reals $\mathbb{R}$ and the circle group $\mathbb{R}/\mathbb{Z}$ have no small subgroups. This is also true for the matrix groups over $\mathbb{R}$.

Combining the preceding results we obtain the following classical results:

2.38 Theorem. If G is a compact [discrete] group, then $G^\wedge$ is a discrete [compact] abelian group.

If X and Y are spaces, then a function $f : X \to Y$ is said to be dense if f(X) is dense in Y, i.e., $\overline{f(X)} = Y$.

Next we state a most important and useful collection of results as a single theorem without proof. Appropriate references are given for each part of the theorem. Recall that the sequence $G \xrightarrow{f} H \xrightarrow{g} K$ of continuous homomorphisms of locally compact abelian groups is exact at H provided $\overline{f(G)} = \ker(g)$.

2.39 Theorem. Let G, H and K be locally compact abelian groups and let $f : G \to K$ be continuous homomorphisms. Then:

(a) $G^\wedge$ is a locally compact abelian group;

(b) $G \overset{\tau}{\approx} (G^\wedge)^\wedge$ under the map Φ defined by $\Phi(x)(\phi) = \phi(x)$ for all $x \in G$ and all $\phi \in G^\wedge$;

(c) If L is a closed subgroup of G, then $L^\wedge \overset{\tau}{\approx} G^\wedge/A$ and $(G/L)^\wedge \overset{\tau}{\approx} A$, where $A = \{\phi \in G^\wedge : \phi(x) = 1$ for all $x \in L\}$;

(d) If $G \xrightarrow{f} H \xrightarrow{g} K$ is exact at H, then $G^\wedge \xleftarrow{f^\wedge} H^\wedge \xleftarrow{g^\wedge} K^\wedge$ is exact at $H^\wedge$;

(e) If $f : G \to H$ is injective [dense], then $f^\wedge : H^\wedge \to G^\wedge$ is dense [injective]. In particular, if G is compact or discrete, then f is injective if and only if $f^\wedge$ is surjective;

(f) If $\{G_i\}_{i\in I}$ is a family of compact abelian groups and $G = \Pi\{G_i\}_{i\in I}$, then $G^\wedge \overset{\tau}{\approx} \Sigma\{G_i^\wedge\}_{i\in I}$;

(g) If $\{G_i\}_{i\in I}$ is a family of discrete abelian groups and $G = \Sigma\{G_i\}_{i\in I}$, then $G^\wedge \overset{\tau}{\approx} \Pi\{G_i^\wedge\}_{i\in I}$;

(h) If $\{G_\alpha, \phi_\alpha^\beta\}_{\alpha\in D}$ is a projective system of compact abelian groups and $G = \varprojlim G_\alpha$, then $\{G_\alpha^\wedge, (\phi_\alpha^\beta)^\wedge\}_{\alpha\in D}$ is an injective system of discrete abelian groups and $G^\wedge \approx \varprojlim G_\alpha^\wedge$; and

(i) If $\{G_\alpha, \phi_\alpha^\beta\}_{\alpha\in D}$ is an injective system of discrete abelian groups and $G = \varprojlim G_\alpha$, then $\{G_\alpha^\wedge, (\phi_\alpha^\beta)^\wedge\}_{\alpha\in D}$ is a projective system of compact abelian groups and $G^\wedge \overset{\tau}{\approx} \varprojlim G_\alpha^\wedge$.

References. For (a), (b), and (c) the reader is referred to either [Pontryagin, 1966] or [Hewitt and Ross, 1963]. Appendix I of [Hofmann and Mostert, 1966] is a good reference for (d) and (e).

For (f) and (g) [Hewitt and Ross, 1963] is again an appropriate source. In fact, this text on Abstract Harmonic Analysis contains a very comprehensive treatment on the entire theory of Pontryagin quality, including a long list of examples. A proof of (h) can be found in [Hurewicz and Wallman, 1959]. Finally, (i) follows from (b), (h), and 2.38.

Verification that the following dual groups are as claimed may be found in [Pontryagin, 1966] or [Hewitt and Ross, 1963]:

(a) $(\mathbb{R}/\mathbb{Z})^\wedge \overset{T}{\approx} \mathbb{Z}$;

(b) $\mathbb{Z}^\wedge \overset{T}{\approx} \mathbb{R}/\mathbb{Z}$; and

(c) $\mathbb{R}^\wedge \overset{T}{\approx} \mathbb{R}$.

We conclude this section with some remarks regarding various topologies one can assign to Hom (S,T), where S and T are topological semigroups with T abelian.

Of all the topologies one could assign to Hom (S,T) the compact-open topology has the unique distinction of being the smallest topology on Hom (S,T) such that the evaluation function $S \in \text{Hom}(S,G) \to T$, $(x,f) \to f(x)$, is (jointly) continuous.

One can assign Hom (S,T) the topology of pointwise convergence or the topology of continuous convergence (see [Kelly, 1955]). If S is locally compact, then the compact-open topology and the topology of continuous convergence agree on Hom (S,T). This fact is a consequence of the following:

2.40 Theorem. Let S be a locally compact semigroup, T an abelian topological semigroup, $\{f_\alpha\}$ a net in Hom (S,T), and $f \in \text{Hom}(S,T)$. Then $\{f_\alpha\}$ converges to f if and only if for each $x \in S$ and each net $\{x_\beta\}$ in S converging to x, the net $\{f_\alpha(x_\beta)\}$ converges to f(x) in T.

Proof. Suppose first that $f = \lim f_\alpha$, and let $\{x_\beta\}$ be a net in S with $x = \lim x_\beta$. We want to show that $f(x) = \lim f_\alpha(x_\beta)$. For this purpose let U be an open subset of T such that $f(x) \in U$ and let V be an open subset of S such that $x \in V$, $\overline{V}$ is compact, and $f(\overline{V}) \subset U$.

Then $f \in N(\overline{V},U)$, so that $\{f_\alpha\}$ is eventually in $N(\overline{V},U)$. We also have $\{x_\beta\}$ is eventually in V, and hence $\{f_\alpha(x_\beta)\}$ is eventually in U. It follows that $\{f_\alpha(x_\beta)\}$ converges to f(x).

Suppose, on the other hand, that $\{f_\alpha(x_\beta)\} \to f(x)$ for each $\{x_\beta\} \to x$ in S. We want to show that $f = \lim f_\alpha$. For this purpose, let K be a compact subset of S and let U be an open subset of T such that $f \in N(K,U)$. We will demonstrate that $\{f_\alpha\}$ is eventually in N(K,U). For the purpose of contradiction, suppose that this is not the case. We can assume that $f_\alpha \in \text{Hom } (S,T)\ N(K,U)$ for each α. Then for each α, there exists $x_\alpha \in K$ such that $f_\alpha(x_\alpha) \in T\backslash U$. In view of the compactness of K the net $\{x_\alpha\}$ has a convergent subnet $\{x_\beta\} \to x$ with $x \in K$, since S is Hausdorff. We have that $f(x) = \lim f_\alpha(x_\beta)$ and $\{f_\alpha(x_\beta)\} \subset T\backslash U$ which is closed. Hence $f(x) \in T\backslash U$, contradicting the fact that $f \in N(K,U)$. ∎

We point out here that the second part of the proof of 2.40 does not employ the fact that S is locally compact. This yields the result that the topology of continuous convergence is larger than the compact-open topology on Hom (S,T) without the assumption that S is locally compact.

> *2.41 Theorem*. Let S and T be topological semigroups with T abelian and let β be a base for the topology of T. Then $\{N(K,U) : K \text{ is a compact subset of } S \text{ and } U \in \mathcal{B}\}$ is a subbase for the topology of Hom (S,T).

Proof. Let C be a compact subset of S, V an open subset of T, and let $f \in N(C,V)$. Then for each $x \in C$, $f(x) \in V$ and there exists $U_x \in \mathcal{B}$ such that $f(x) \in U_x \subset V$. Since f is continuous, for each $x \in C$, there exists an open set V_x in S such that $x \in U_x$ and $f(\overline{V}_x \cap C) \subset U_x$. Let $\{V_i : i = 1,\ldots,n\}$ be a finite subcollection of $\{V_x : x \in C\}$ covering C and let $\{U_i : i = 1,\ldots,n\}$ be the corresponding subcollection of $\{U_x : x \in C\}$. Then $f \in \bigcap_{j=1}^{n} (\overline{V} \cap C, U_j) \subset N(C,V)$. ∎

SEMIGROUP COMPACTIFICATIONS

Associated with each topological semigroup S, there is a compact semigroup called the Bohr compactification of S which is universal over the compact semigroups containing dense continuous homomorphic images of S. In this section we give a proof of the existence and uniqueness of Bohr compactifications, a brief discussion of other types of semigroup compactifications, an elementary proof of the product theorem, a result which states that a topological semigroup with identity has the same Bohr compactification as a dense ideal, an application of this result to the theory of compact semigroups, examples of Bohr compactifications, and a well-known result which gives a method for computing the Bohr compactification of a locally compact abelian group. Bohr (or almost periodic) compactifications of semigroups have been studied in [DeLeuuw and Glicksberg, 1961], [Berglund and Hofmann, 1967], and [Anderson and Hunter, 1969]. The reader is referred to [Hewitt and Ross, 1963] for a discussion of Bohr compactifications of locally compact abelian groups.

If S is a topological semigroup, then a *Bohr compactification* of S is a pair (β, B) such that B is a compact semigroup, $\beta : S \to B$ is a continuous homomorphism, and if $g : S \to T$ is a continuous homomorphism of S into a compact semigroup T, then there exists a unique continuous homomorphism $f : B \to T$ such that the diagram:

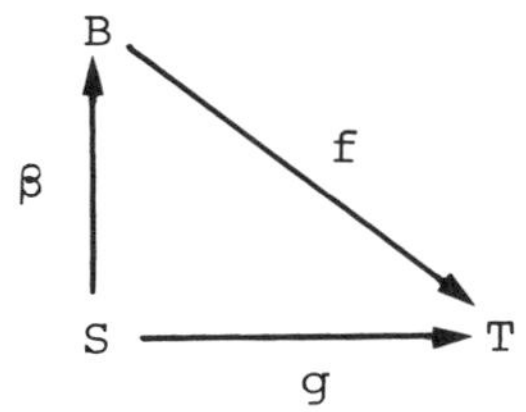

commutes.

We will employ two lemmas to prove that each topological semigroup has a Bohr compactification.

If X is a set, then Card X denotes the cardinal number of X.

2.42 Lemma. Let D be a dense subset of a Hausdorff space X. Then Card $X \leqslant 2^{2^D}$.

Proof. The function $f : X \to 2^{2^D}$ defined by $f(x) = \{U : U = V \cap D$ for some neighborhood V of x in X$\}$ is an injection. ∎

Suppose that T is a topological semigroup and $f : T \to K$ is a bijection from T onto a set K. Then, as we have seen, we can give K the quotient topology τ induced by f. In addition to this, we can give K the multiplication m defined to be the composition $K \times K \xrightarrow{f^{-1} \times f^{-1}} T \times T \xrightarrow{p} T \xrightarrow{f} K$, where p is the multiplication on T. With the topology τ and multiplication m, K becomes a topological semigroup and $f : T \to K$ is a topological isomorphism. We call m *the multiplication on* K *induced by* f.

2.43 Lemma. Let S be a topological semigroup. Then there exists a collection $\{(\phi_\alpha, S_\alpha) : \alpha \in A\}$ such that S_α is a compact semigroup and $\phi_\alpha : S \to S_\alpha$ is a dense continuous homomorphism for each $\alpha \in A$, and if $g : S \to T$ is a dense continuous homomorphism of S into a compact semigroup T, then there exists $\delta \in A$ and a topological isomorphism $f : S_\delta \to T$ such that the following diagram commutes:

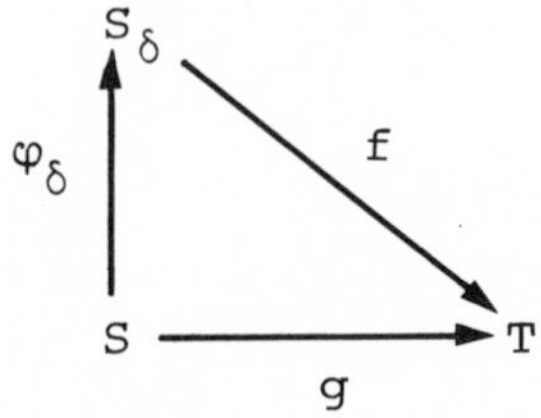

Proof. Let Λ denote the family of all quadruples (K,τ,m,σ) such that $K \subset 2^{2^S}$,   is a compact Hausdorff topology on K, m is a continuous associative multiplication on K, and $\sigma : S \to K$ is a dense homomorphism. Observe that Λ is non-empty, since each singleton of 2^{2^S} admits the conditions required to be a member of Λ. A straightforward cardinality argument yields that Λ is a set. We will show that Λ is the desired collection. For this purpose, suppose that $g : S \to T$ is a dense continuous homomorphism of S into a compact semigroup T. In view of 2.42, we see that Card $T \leqslant 2^{2^{g(S)}} \leqslant$ Card 2^{2^S}. It follows that there exists $K \subset 2^{2^S}$ and a bijection $k : T \to K$. Let τ be the quotient topology on K induced by k, m the multiplication on K induced by k, and define $\sigma : S \to K$ by $\sigma = k \circ g$. Then $(K,\tau,m,\sigma) \in \Lambda$ and $k^{-1} : K \to T$ is a topological isomorphism such that $k^{-1} \circ \sigma = g$. ■

2.44 Theorem. (Existence of Bohr Compactifications). If S is a topological semigroup, then there exists a Bohr compactification (β,B) of S.

Proof. Let $\{(\phi_\alpha,S_\alpha) : \alpha \in A\}$ be the collection of 2.43 and define $\sigma : S \to \Pi\{S_\alpha\}_{\alpha\in A}$ so that $\sigma(x)(\alpha) = \phi_\alpha(x)$ for each $\alpha \in A$ and $x \in S$. Then, clearly σ is a continuous homomorphism. Let $B = \overline{\sigma(S)}$ and define $\beta : S \to B$ so that $\beta(x) = \sigma(x)$ for each $x \in S$. We will show that (β,B) is a Bohr compactification of S. For this purpose, suppose that $g : S \to T$ is a continuous homomorphism of S into a compact semigroup T. With no loss of generality, we can assume that g is dense. Then, by 2.43, there exists $\delta \in A$ and a topological isomorphism $h : S_\delta \to T$ such that $h \circ \phi = g$. Define $f : B \to T$ by $f = h \circ \pi_\delta$. Then $f \circ \beta = g$, and f is a continuous homomorphism. The uniqueness of f is a consequence of the fact that $\beta : S \to B$ is dense. ■

2.45 Theorem. (Uniqueness of Bohr Compactifications). Let S be a topological semigroup, and let (α,A) and (β,B) be Bohr compactifications of S. Then there exists a topological isomorphism $\psi : A \to B$ such that the diagram:

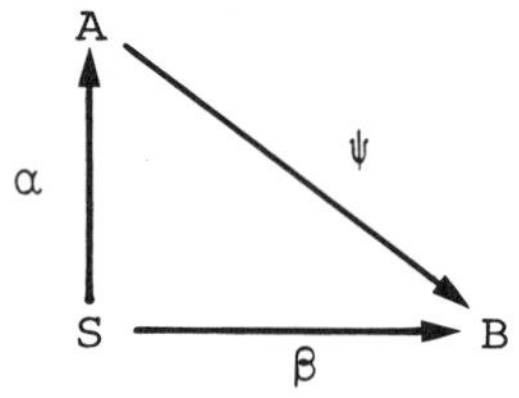

commutes.

Proof. Since (α, A) is a Bohr compactification of S, there exists a continuous homomorphism $\psi : A \to B$ such that the diagram commutes, and likewise, since (β, B) is a Bohr compactification of S, there exists a continuous homomorphism $\phi : B \to A$ such that $\alpha = \phi \circ \beta$. In view of the commuting diagrams:

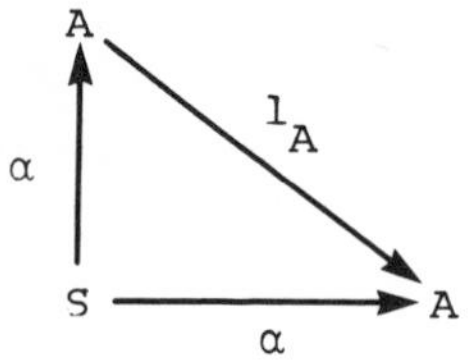

and

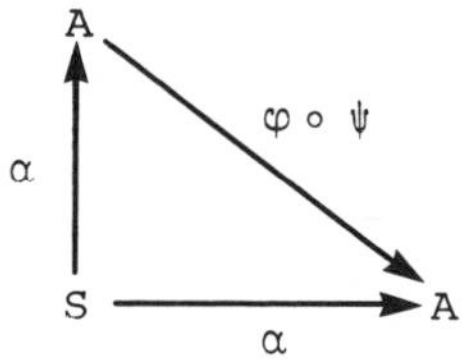

and the uniqueness of $\phi \circ \psi$, we see that $\phi \circ \psi = 1_A$, and similarly $\psi \circ \phi = 1_B$. We conclude that ψ is a topological isomorphism. ∎

Observe that if S is a compact semigroup, then $(1_S, S)$ is the Bohr compactification of S.

If we consider two topological semigroups S and T and their respective Bohr compactifications (α, A) and (β, B), then each continuous homomorphism $g : S \to T$ gives rise to a unique continuous homomorphism $f : A \to B$ such that the diagram:

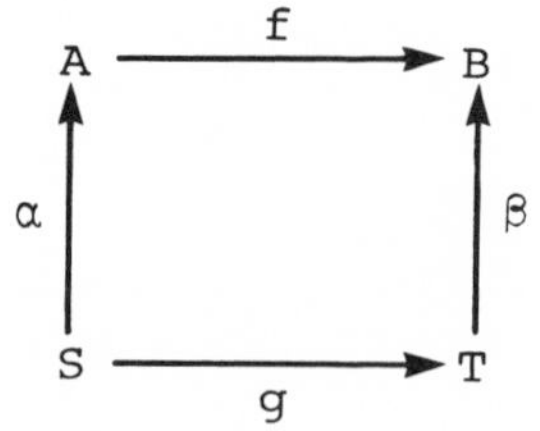

commutes.

One should observe the analogy between the Borh compactification of a topological semigroup and the generalized Stone-Cech compactification of a Hausdorff space. In fact, the Stone-Cech compactification can be employed to establish the existence of the Bohr compactification of a topological semigroup [Stepp, 1974]. Suppose that S is a topological semigroup and $\gamma : S \to C$ is the Stone-Cech compactification of S as a Hausdorff space. Let $\{(R_\alpha, m_\alpha) : \alpha \in A\}$ denote the collection of all pairs such that R_α is a closed equivalence on C, m_α is a continuous associative multiplication on C/R_α, and $\pi_\alpha \circ \gamma : S \to C/R_\alpha$ is a homomorphism, where $\pi_\alpha : C \to C/R_\alpha$ is the natural map for each $\alpha \in A$. This collection is clearly a non-empty set, since $C \times C$ with trivial multiplication on the quotient space is a member. Let $R = \cap \{R_\alpha\}_{\alpha \in A}$ and $\pi : C \to C/R$ the natural map. It is straightforward to show that C/R admits a continuous associative multiplication so that $(\pi \circ \gamma, C/R)$ is the Bohr compactification of S.

There are other types of compactifications one could consider for a given topological semigroup S, such as the group compactification. In the case S is an abelian topological semigroup and (β, B) is its Bohr compactification, it can easily be established that $(\sigma \circ \beta, M(B))$ is the group compactification of S, where $\sigma : B \to M(B)$ is defined by $\sigma(x) = xe$, and e is the identity of the compact abelian group M(B).

Another type of compactification (which generally does not have a universal mapping property as does the Bohr compactification) is the one-point compactification of a locally compact semigroup. We have witnessed this compactification in previous examples, such as $\mathbb{H}^* = \mathbb{H} \cup \{\infty\}$ and $\mathbb{N}^* = \mathbb{N} \cup \{\infty\}$. In some situations the ideal point may be a zero (as in $\mathbb{H}^*$ and $\mathbb{N}^*$), in others an identity, in the others neither of these, and in still other situations it may not be possible to continuously extend the multiplication on the semigroup to the ideal point. Before considering examples of these situations we prove a result which will be useful in these illustrations.

2.46 Theorem. Let S be a dense locally compact subsemigroup of a compact semigroup T such that the multiplication on S can be extended continuously to $S \cup \{\infty\}$. If T has an identity [zero] in T\S, then ∞ is an identity [zero] for $S \cup \{\infty\}$.

Proof. Suppose that 1 is an identity for T in T\S. Let $\{a_\alpha\}$ be a net in S converging to 1. Then $\{a_\alpha\}$ has a subnet in $S \cup \{\infty\}$ converging to ∞. We can assume that $\{a_\alpha\} \to \infty$ in $S \cup \{\infty\}$. Fix $x \in S$. Then $\{a_\alpha x\} \to x$ in S since $\{a_\alpha\} \to 1$ in T, and hence $\infty x = x$ and similarly $x\infty = x$. We obtain that ∞ acts as an identity on elements of S. Since S is dense in $S \cup \{\infty\}$, we have that ∞ is an identity for $S \cup \{\infty\}$. A similar argument works for a zero in T\S. ∎

We present some examples.

If $S = [0,1)$ in I_u (the usual interval), then by defining $x\infty = \infty x = x$ for each $x \in S \cup \{\infty\}$ we obtain that $S \cup \{\infty\}$ is a compact semigroup which is topologically isomorphic to I_u, so that ∞ is an identity for $S \cup \{\infty\}$.

If $S = (0,1]$ in I_u, then by defining $x\infty = \infty x = x$ for each $x \in S \cup \{\infty\}$, we see that $S \cup \{\infty\} \overset{T}{\approx} I_u$, and ∞ is a zero for $S \cup \{\infty\}$.

If $S = [0,\frac{1}{2})$ in I_u, then by defining $\infty x = x\infty = \frac{1}{2}x$ for $x \in S$ and $\infty\infty = \frac{1}{4}$, we obtain that $S \cup \{\infty\} \overset{T}{\approx} [0,\frac{1}{2}]$, so that ∞ is neither an identity or a zero for $S \cup \{\infty\}$.

If $S = (0,1)$ in I_u, then in view of 2.46 we see that the multiplication on S cannot be continuously extended to $S \cup \{\infty\}$, since otherwise ∞ would serve as both a zero and an identity for $S \cup \{\infty\}$ (which is clearly impossible).

We turn again to consideration of the Bohr compactification.

One of the problems arising in working with the Bohr compactification (β,B) of a topological semigroup S is that $\beta : S \to B$ is not always injective. It is trivial [not difficult] to show that there is a continuous injective homomorphism [embedding] $g : S \to T$ of S into a compact semigroup T if and only if $\beta : S \to T$ is injective [an embedding]. A set of sufficient conditions for β to be injective was established in [Hildebrant and Lawson, 1972] and generalized in

[Friedberg, 1975]. The following example illustrates that β is not always injective.

Let (β, $\mathbb{B}$) be the Bohr compactification of the discrete bicyclic semigroup $\mathbb{B}$. Let e = β(1). Then, since β is dense, e is an identity for B. Suppose that β is injective. Then $\beta(x)\beta(y) \in H(e)$, where x = (0,1) and y = (1,0), but neither β(x) nor β(y) is in H(e) contradicting 1.17. We conclude that β is not injective, and hence, by the above remark, no compact semigroup contains an injective homomorphic image of $\mathbb{B}$.

The following is an example of topological semigroup S with Bohr compactification (β,B) such that β : S → B is a continuous injective homomorphism but β is not a topological embedding.

Let S be the group $\mathbb{R}/\mathbb{Z}$ with the discrete topology and (β,B) the Bohr compactification of S. Let π : R → $\mathbb{R}/\mathbb{Z}$ denote the natural map, b an irrational in $\mathbb{R}$, and a = π(b) in S. Then, since B is compact, the sequence $\{\beta(a^n)\}_{n\in\mathbb{N}}$ clusters in B. However, the sequence $\{a^n\}_{n\in\mathbb{N}}$ does not cluster in S, proving that β^{-1} is not continuous. It is clear that β is a continuous injective homomorphism, since the identity function S → $\mathbb{R}/\mathbb{Z}$ is a continuous injective homomorphism.

A second type of problem occurring in the study of Bohr compactifications is that of computing this compactification for a specific topological semigroup. This computation was performed for $\mathbb{N}$ and $\mathbb{H}$ in [Hofmann and Mostert, 1966] and for the positive discrete rationals in [Hildebrant, 1969]. One of the more useful tools in computing Bohr compactifications is the Product Theorem. Here the distinction between the Bohr compactification and its topological analog the Stone-Cech compactification is more pronounced, since the Stone-Cech compactification does not generally have the product property even for a finite number of factors. In proving the Product Theorem, we will employ the following lemma.

<u>*2.47 Lemma*</u>. Let S be a compact semigroup with identity 1 and let a and b be distinct points of S. Then there exist open sets M, N and V in S with a ∈ m, b ∈ N, 1 ∈ V, and M ∩ N = □ such that for each p ∈ S, there exists an open

set G_p containing p so that either $G_p V \cap M = \square$ or $G_p V \cap N = \square$.

Proof. Let M and N be open sets such that $a \in M$, $b \in N$, and $\overline{M} \cap \overline{N} = \square$. Then $\Delta \subset \mathcal{U} \subset \mathcal{V}$, where $\mathcal{V} = (S \times S)\backslash(\overline{M} \times \overline{N})$, for some Δ-ideal $\mathcal{U}$ (see 1.42). Let V be an open set such that $1 \in V$ and $\overline{V} \times \overline{V} \subset \mathcal{U}$. Now for $p \in S$, we have $(p,p)(\overline{V} \times \overline{V}) \subset \mathcal{U}$, since $\mathcal{U}$ is a Δ-ideal. Hence, by 1.1, there exists an open set G_p containing p with $(G_p \times G_p)(\overline{V} \times \overline{V}) \subset \mathcal{U} \subset \mathcal{V}$. It follows that either $G_p\overline{V} \cap M = \square$ or $G_p\overline{V} \cap N = \square$. ■

The following result can be found in [DeLeuuw and Glicksberg, 1961].

2.48 The Product Theorem. Let $\{S_\alpha : \alpha \in A$ be a collection of abelian topological monoids, (β_α, B_α) the Bohr compactification of S_α for each $\alpha \in A$, and β : $\Pi\{S_\alpha\}_{\alpha\in A} \longrightarrow \Pi\{\beta_\alpha\}_{\alpha\in A}$ the function defined by $\beta(x)(\delta) = \beta_\delta\pi_\delta(x)$, where $\pi_\delta : \Pi\{S_\alpha\}_{\alpha\in A} \longrightarrow S$ is projection for each $\delta \in A$. Then $(\beta, \Pi\{B_\alpha\}_{\alpha\in A})$ is the Bohr compactification of $\Pi\{S_\alpha\}_{\alpha\in A}$.

Proof. A straightforward argument works to demonstrate that β is a dense continuous homomorphism. Let $S = \Pi\{S_\alpha\}_{\alpha\in A}$, $B = \Pi\{B_\alpha\}_{\alpha\in A}$, and suppose that $g : S \to T$ is a continuous homomorphism of S into a compact semigroup T. With no loss of generality we can assume that g is dense. To complete the proof, we need to exhibit a continuous homomorphism $f : B \to T$ such that the diagram:

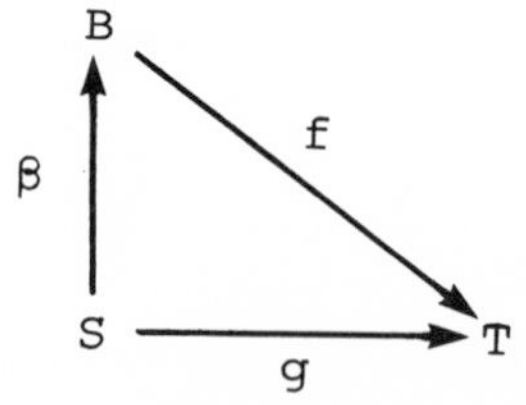

commutes. Let 1 denote the identity of S and let 1_α denote the identity of S_α and $p_\alpha : B \to B_\alpha$ projection for each $\alpha \in A$. Define

$\psi_\alpha : S_\alpha \to S$ so that

$$\pi_\delta\psi_\alpha(z) = \begin{cases} z & \text{if } \alpha = \delta; \\ 1_\delta & \text{if } \alpha \neq \delta, \end{cases}$$

for each $\alpha, \delta \in A$ and $z \in S_\alpha$. Then $\psi_\alpha : S_\alpha \to S$ is a continuous homomorphism for each $\alpha \in A$, and there exists a unique continuous homomorphism $g_\alpha : B_\alpha \to T$ such that the diagram:

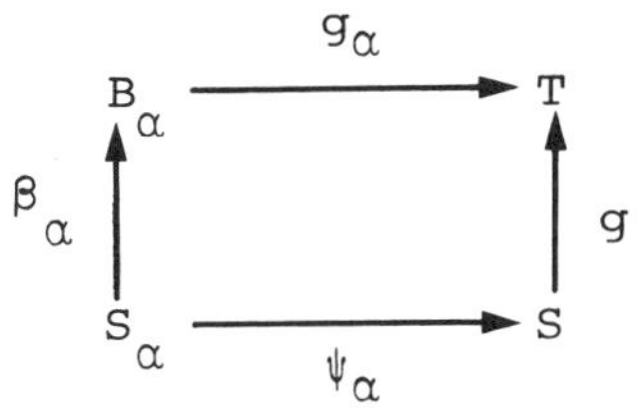

commutes, since (β_α, B_α) is the Bohr compactification of S_α for each $\alpha \in A$.

Now, in order to define $f : B \to T$, let $x \in B$ and let $\mathcal{U}$ be the collection of all open neighborhoods of x. Define $G(x) = \cap\, \{g\beta^{-1}(U)\}_{U \in \mathcal{U}}$. We will show that G(x) is degenerate and define f(x) to be that point. Since β is dense, we have that $\{g\beta^{-1}(U) : U \in \mathcal{U}\}$ is a descending collection of non-empty compact subsets of T, and hence $G(x) \neq \square$.

For the purpose of demonstrating that G(x) is degenerate, suppose that G(x) contains distinct points a and b, and select M, N and V in T according to 2.47 with $a \in M$, $b \in N$, and $g(1) \in V$ (observe that g(1) is the identity of T). Let W be a basic open subset of S containing 1 such that $g(W) \subset V$. Then there exists a finite subset F of A such that $\pi_\alpha(W) = S_\alpha$ for $\alpha \in A\backslash F$. Let p be the product of the elements of the finite set $\{g_\alpha p_\alpha(x) : \alpha \in F\}$, and choose G_p according to 2.47. Without loss of generality we assume that $G_pV \cap M = \square$. In view of the continuity of multiplication in T and the continuity of $g_\alpha : B_\alpha \to T$ for each $\alpha \in A$, we see that for each $\alpha \in F$, there exists an open subset H_α of B_α containing $p_\alpha(x)$ such that the product of the set $\{g_\alpha(H_\alpha) : \alpha \in F\}$ is contained in G_p. Let H be the basic open subset of B containing x such that $\beta_\alpha(H) = H_\alpha$ for $\alpha \in F$ and $p_\alpha(H) = B_\alpha$ for $\alpha \in A\backslash F$. Then there exists $s \in S$ such that $g(s) \in M$ and $\beta(s) \in H$. Now let t and W be the elements of S such that for each $\alpha \in A$,

$$\pi_\alpha(t) = \begin{cases} \pi_\alpha(S) & \text{if } \alpha \in F; \\ 1_\alpha & \text{if } \alpha \in A\backslash F, \end{cases}$$

and

$$\pi_\alpha(W) = \begin{cases} 1_\alpha & \text{if } \alpha \in F; \\ \pi_\alpha(S) & \text{if } \alpha \in A\backslash F. \end{cases}$$

Then $s = tw$, $\beta(t) \in H$, and $w \in W$. We obtain $g(t) \in G_p$, and hence $g(s) = g(tw) = g(t)g(w) \in G_pV$. On the other hand, we have that $g(s) \in M$ contradicting the fact that $G_pV \cap M = \square$, and we conclude that G(x) is degenerate. We define f(x) to be the point in G(x) for each $x \in B$. Then $g = f\beta$ is immediate.

To establish the continuity of f, let $x \in B$ and let Q be an open subset of T containing f(x). Since $\{g\beta^{-1}(U) : U \in \mathcal{U}\}$ is descending, there exists $U \in \mathcal{U}$ such that $g\beta^{-1}(U) \subset Q$, and thus $f(U) \subset Q$. It follows that f is continuous. Since $g = f\beta$, $f|\beta(S)$ is a homomorphism, and g is dense, we conclude that f is a homomorphism. ■

We turn to the proof of a result on Bohr compactifications which has an important application (2.54) to the theory of compact semigroups.

A net $\{a_\alpha\}$ in a topological semigroup S is said to be an *approximate identity* provided $\{xa_\alpha\} \to x$ and $\{a_\alpha x\} \to x$ for each $x \in S$.

2.49 Lemma. Let S be a topological semigroup with an approximate identity $\{a_\alpha\}$ and let (β,B) denote the compactification of S. Then B has an identity e and $\{\beta(a_\alpha)\}$ converges to e.

Proof. Since B is compact, the net $\{\beta(a_\alpha)\}$ clusters to some $e \in B$. Let $x \in S$. Then $\{a_\alpha x\} \to x$ in S, so that $\{\beta(a_\alpha)\beta(x)\} \to \beta(x)$ in B. Since $\{\beta(a_\alpha)\}$ clusters to e, we have that $e\beta(x) = \beta(x)$ for each $x \in S$. It follows that e is a left identity for $\beta(S)$, and since $\overline{\beta(S)} = B$, is a left identity for B. Similarly, e is a right identity for B. ■

2.50 Corollary. Let T be a dense subsemigroup of a topological monoid S and let (β,B) denote the Bohr compactification of T. Then B has an identity.

Proof. Let $\{a_\alpha\}$ be a net in T converging to the identity of S. Then $\{a_\alpha\}$ is an approximate identity for T, and the corollary follows from 2.49. ■

The next result appears in [Hildebrant and Lawson, 1973].

2.51 Theorem. Let S be a topological monoid with Bohr compactification (β,B) and let I be a dense ideal of S with Bohr compactification (α,A). Then there exists a topological isomorphism $\phi : A \to B$ such that the diagram:

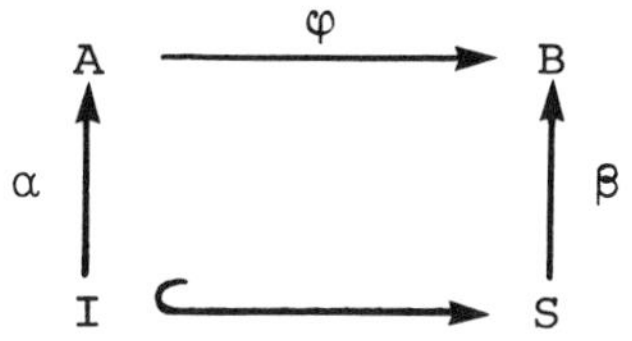

commutes.

Proof. In view of 2.50, we see that A has an identity e. Since (α,A) is the Bohr compactification of S, we obtain a continuous homomorphism $\phi : A \to B$ such that the above diagram commutes. It remains to show that ϕ is a topological isomorphism.

Let $x \in S$ and let $\{x_\lambda\}$ and $\{y_\lambda\}$ be nets in I converging to s. Then $\{\alpha(x_\beta)\}$ and $\{\alpha(y_\gamma)\}$ cluster to points p and q, respectively, in A. We will show that $p = q$. By considering subnets, we can assume that $\{\alpha(x_\lambda)\} \to p$ and $\{\alpha(y_\gamma)\} \to q$. Let $k \in I$. Then $\{kx_\lambda\} \to kx$ and $\{ky_\gamma\} \to kx$, and hence $\{\alpha(kx_\lambda)\} \to \alpha(kx)$. We obtain that

$\{\alpha(kx_\lambda)\} = \{\alpha(k)\alpha(x_\lambda)\} \to \alpha(k)p$, so that $\alpha(kx) = \alpha(k)p$, and similarly, $\alpha(kx) = \alpha(k)q$. It follows that $\alpha(k)p = \alpha(k)q$ for each $k \in I$. Since α is dense, we have that $p = ep = eq = q$.

Define $\sigma : S \to A$ by $\sigma(x) = \lim \alpha(x_\gamma)$, where $\{x_\gamma\}$ is a net in I converging to x. It is straightforward to prove that $\sigma : S \to A$ is a continuous homomorphism by employing the fact $\sigma|I = \alpha$. In view of the fact that (β,B) is the Bohr compactification of S, we see that there exists a continuous homomorphism $\psi : B \to A$ such that the diagram:

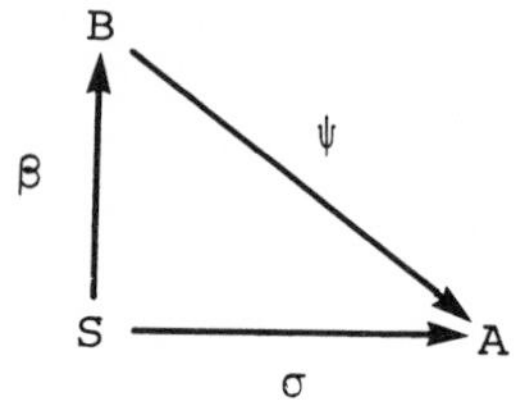

commutes. Since $\sigma|I = \alpha$, we have that $\psi \circ \phi = 1_A$ and $\phi \circ \psi = 1_B$, so that ϕ is a topological isomorphism. ■

2.52 Corollary. If S is a compact monoid and I is a dense ideal of S, then (j,S) is the Bohr compactification of I, where $j : I \hookrightarrow S$.

2.53 Corollary. Let S be a locally compact non-compact semigroup such that the multiplication on S can be continuously extended to the one-point compactification $S \cup \{\infty\}$ with ∞ as an identity. Then $(j, S \cup \{\infty\})$ is the Bohr compactification of S, where $j : S \hookrightarrow S \cup \{\infty\}$.

2.54 Corollary. Let S be a compact connected semigroup with identity 1. Then S is uniquely determined by $S \backslash H(1)$.

If we want an intuitive notion of the structure of a monoid S, then we think of the group of units H(1) as floating on the top and the minimal ideal M(S) as sitting at the bottom (or base) of S, since multiplication "flows down" toward M(S). We have observed that M(S) can be replaced by a continuous surmorphic image of M(S);

thus replacing the base in case S is abelian. Observe that one implication of 2.54 is that the top cannot be replaced when S is connected.

In [Hildebrant and Lawson, 1973] the result of 2.51 was established in a slightly more general setting. The requirement that S have an identity can be replaced by the condition that S = ESE or that I is divisible, i.e., each element of I has an nth root for each $n \in \mathbb{N}$.

In later sections we will compute the Bohr compactification of $\mathbb{N}$ and $\mathbb{H}$. The result of 2.51 allows us to present the following simple example:

Let $S = I_u \times I_u$ and $I = [0,1) \times [0,1)$. Then, in view of 2.52, we see that (j,S) is the Bohr compactification of I, where $j : I \hookrightarrow S$. Observe that $[0,1) \hookrightarrow I_u$ is the Bohr compactification of [0,1), so that the existence of an identity in the Product Theorem is not a necessary condition.

The final result of this section gives a method for computing the Bohr compactification of a locally compact abelian group, and is presented without proof (see [Hewitt and Ross, 1963]).

2.55 Theorem. Let G be a locally compact abelian group and let $(G^\wedge)_d$ be the character group of G given the discrete topology. Define $\Phi : G \to (G^\wedge)_d^\wedge$ by $\Phi(x)(\phi) = \phi(x)$ for all $x \in G$ and $\phi \in (G^\wedge)_d$. Then $(\Phi, (G^\wedge)_d^\wedge)$ is the Bohr compactification of G.

Recall that a topological group G is monothetic provided there is a dense continuous homomorphism $\phi : \mathbb{Z} \to G$.

Consider the locally compact abelian group $\mathbb{Z}$. Now that $\mathbb{Z}^\wedge \overset{T}{\approx} \mathbb{R}/\mathbb{Z}$. We conclude that the object of the Bohr compactification of $\mathbb{Z}$ is the character group of the discrete group $(\mathbb{R}/\mathbb{Z})_d^\wedge$. Thus any compact monothetic group is a continuous homomorphic image of $(\mathbb{R}/\mathbb{Z})_d^\wedge$. For this reason, the group $(\mathbb{R}/\mathbb{Z})_d^\wedge$ is called *the universal compact monothetic group*.

A topological group G is said to be solenoidal if there exists a dense continuous homomorphic $\phi : \mathbb{R} \to G$.

Consider the locally compact abelian group $\mathbb{R}$. We have seen that $\mathbb{R}^\wedge \overset{\tau}{\approx} \mathbb{R}$, and hence the object of the Bohr compactification of $\mathbb{R}$ is $\mathbb{R}_d^\wedge$. It follows that any compact solenoidal group is a continuous homomorphic image of $\mathbb{R}_d^\wedge$. For this reason, the group $\mathbb{R}_d^\wedge$ is called *the universal compact solenoidal group.*

FREE TOPOLOGICAL SEMIGROUPS

In the previous sections of this chapter we have produced new topological semigroups from a given topological semigroup or a given collection of topological semigroups. Here we deviate from this pattern slightly by starting with a space X and producing a topological semigroup called the free topological semigroup generated by X which has a universal mapping property relative to X. Following a proof of existence and uniqueness of free topological semigroups we present a characterization which parallels the algebraic theory in [Clifford and Preston, 1961], where one begins with a set and produces the free algebraic semigroup generated by the given set. Free topological semigroups have been studied in [Christoph, 1970].

If X is a space, then a *free topological semigroup* generated by X is a pair (ϕ, F) such that F is a topological semigroup, $\phi : X \to F$ is a continuous function, and if $g : X \to T$ is a continuous function from X into a topological semigroup T, then there exists a unique continuous homomorphism $f : F \to T$ such that the diagram:

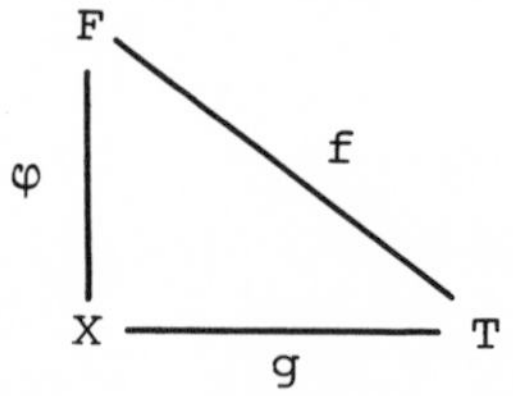

commutes.

2.56 Lemma. Let T be a topological semigroup and D a subset of T such that $T = \theta(D)$. Then Card $T \leqslant$ Card $\cup \{D^n : n \in \mathbb{N}\}$.

Proof. This is immediate from the fact that each element of T is a finite product of elements of D. ■

2.57 Lemma. Let X be a space. Then there exists a collection $\{(\phi_\alpha, S_\alpha) : \alpha \in A\}$ such that S is a topological semigroup, $\phi_\alpha : X \to S_\alpha$ is a continuous function with $S_\alpha = \theta(\phi_\alpha(X))$ for each $\alpha \in A$, and if $g : X \to T$ is a continuous function from X into a topological semigroup T such that $T = \theta(g(X))$, then there exists $\delta \in A$ and a topological isomorphism $f : S_\delta \to T$ such that the diagram:

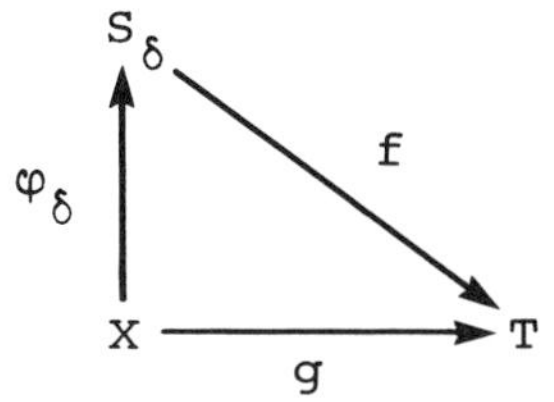

commutes.

Proof. Let $Y = \{X^n : n \in \mathbb{N}\}$, and let A denote the collection of all quadruples (S,τ,m,σ) such that $S \subset 2^{2^{X^{\mathbb{N} \cup Y}}}$, τ is a Hausdorff topology on S, m is a continuous associative multiplication on the space (S,τ), and $\sigma : X \to S$ is a continuous function such that $S = \theta(\sigma(X))$. Then Λ is non-empty, since each singleton of $2^{2^{X^{\mathbb{N} \cup Y}}}$ admits the conditions required to be a member of Λ. A straightforward cardinality argument yields that Λ is a set. We will show that Λ is the desired collection. For this purpose, suppose that $g : X \to T$ is a continuous function from X into a topological semigroup T with $T = \theta(g(X))$. Then, by virtue of 2.56, we have that Card $T \leqslant$ Card $2^{2^{\theta(t(X))}} \theta(g(X)) \leqslant$ Card $2^{2^{g(X)^{\mathbb{N} \cup Y}}}$. It follows that there exists $S \subset 2^{2^{X^{\mathbb{N} \cup Y}}}$ and a bijection $k : T \to S$. Using 2.42, with k as the bijection, a topology τ and a multiplication m are induced on S. Define $\sigma : X \to S$ by $\sigma = k \circ g$. Then $(S,\tau,m,\sigma) \in \Lambda$ and

$k^{-1} : S \to T$ is a topological isomorphism such that the diagram:

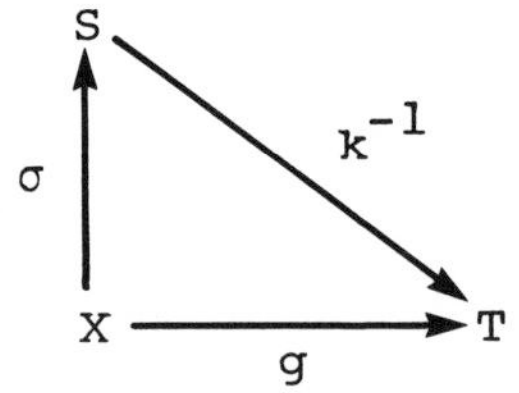

commutes. ■

2.58 Theorem. (Existence of Free Topological Semigroups). If X is a space, then there exists a free topological semigroup (ϕ, F) generated by X.

Proof. Let $\{(\phi_\alpha, S_\alpha) : \alpha \in A\}$ be the collection of 2.57. Define $\sigma : X \to \Pi\{(S_\alpha : \alpha \in A\}$ so that $\sigma(x)(\alpha) = \phi_\alpha(x)$ for each $\alpha \in A$ and $x \in X$. Then σ is a continuous function, since the composition of σ with each projection is continuous. Let $F = \theta(\sigma(X))$ in $\Pi\{S_\alpha : \alpha \in A\}$, and define $\phi : X \to F$ so that $\phi(x) = \sigma(x)$ for each $x \in X$. We will show that (ϕ, F) is a free topological semigroup generated by X. For this purpose, suppose that $g : X \to T$ is a continuous function from X into T. With no loss of generality, we can assume that $T = \theta(g(X))$. Then, by 2.57, there exists $\delta \in A$ and a topological isomorphism $h : S_\delta \to T$ such that $h \circ \phi_\delta = g$. Define $f : F \to T$ by $f = h \circ \pi_\delta$. Then the diagram:

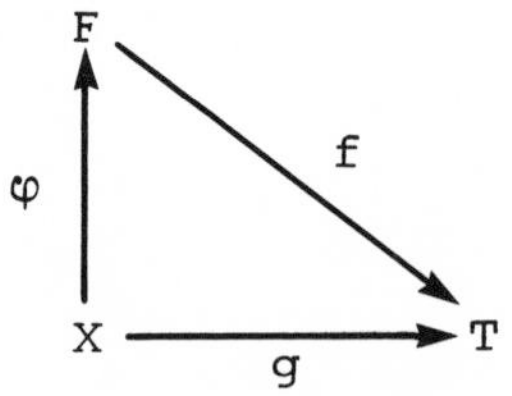

commutes, and f is a continuous homomorphism. The uniqueness of f is a consequence of the fact that $F = \theta(\phi(X))$. ■

2.59 Theorem. (Uniqueness of Free Topological Semigroups). Let X be a space, and (α, A) and (β, B) free topological semigroups generated by X. Then there exists a topological isomorphism $\psi : A \to B$ such that the diagram:

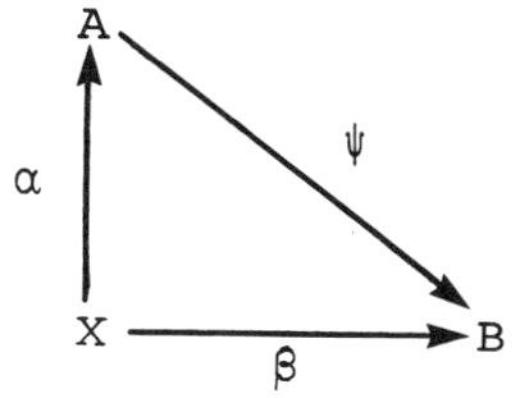

commutes.

Proof. Since (α, A) is a free topological semigroup generated by X, there is a continuous homomorphism $\psi : A \to B$ such that the diagram above commutes, and likewise a continuous homomorphism $\phi : B \to A$ such that $\alpha = \phi \circ \beta$. In view of the commuting diagrams:

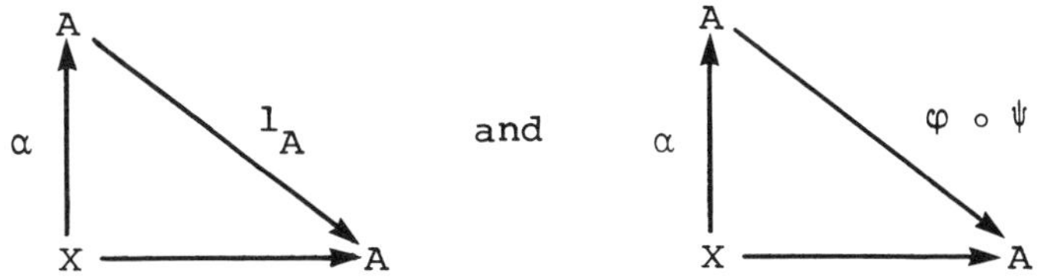

and the uniqueness of diagram completing homomorphisms, we see that $\phi \circ \psi = 1_A$, and similarly $\psi \circ \phi = 1_B$. We conclude that ψ is a topological isomorphism. ■

If we consider two spaces X and Y and their respective free topological semigroups (ϕ, S) and (ψ, T), then it is clear that each continuous function $f : X \to Y$ gives rise to a unique continuous homomorphism $g : S \to T$ such that the diagram:

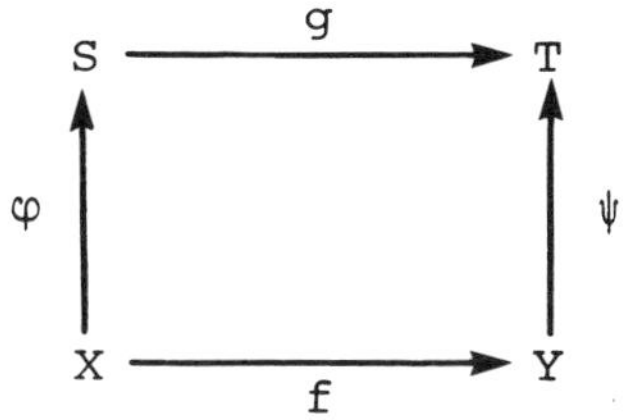

commutes.

Let X be a space and let (ϕ, F) be the free topological semigroup generated by X. If $G = \theta(\phi(X))$, ψ is the obvious restriction of ϕ, g is the map obtained from $\phi : X \to G$, and h is the inclusion of G into G, then the diagram:

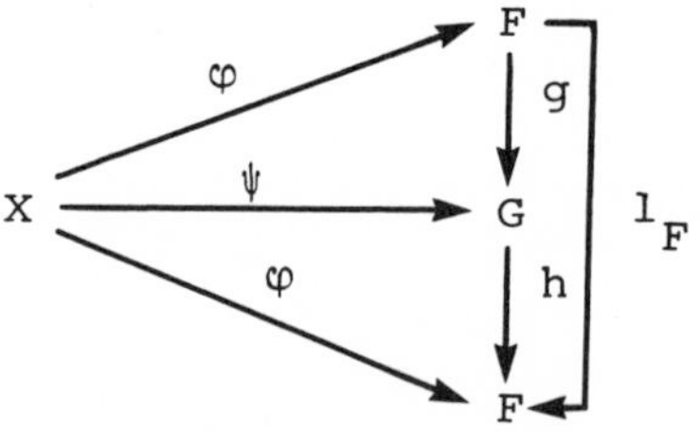

implies h is surjective and so F = G. Hence $\phi(X)$ algebraically generates F. In the case that X is Hausdorff we obtain that $\phi : X \to F$ is a topological embedding by employing the following characterization of the free topological semigroup generated by a Hausdorff space:

> <u>*2.60 Theorem.*</u> Let X be a Hausdorff space and F the set of all finite sequences of elements of X. Define multiplication on F by $((x_1,\ldots,x_n),(y_1,\ldots,y_n)) \longmapsto (x_1,\ldots,x_n, y_1,\ldots,y_m)$, define $\phi : X \to F$ by $\phi(x) = (x)$ for each $x \in X$, and let $\mathcal{B}$ denote the set of all finite cartesian products of open subsets of X. Then $\mathcal{B}$ is a base for a topology on F, relative to which (ϕ, F) is the free topological semigroup generated by X.

Proof. It is straightforward to show that F is a topological semigroup, $\phi : X \to F$ is continuous, and that $F = \theta(\phi(X))$. To complete the proof, suppose that $g : X \to T$ is a continuous function from X into a topological semigroup T. Define $f : F \to T$ by $f((x_1,\ldots,x_n)) = g(x_1)\cdots g(x_n)$. It is clear that f is a homomorphism and that $f \circ \phi = g$. To see that f is continuous, let $(a_1,\ldots,a_m) \in F$ and W be an open subset of T containing $f((a_1,\ldots,a_m)) = g(a_1)\cdots g(a_m)$. In view of the continuity of multiplication in T, we see that there exist open sets $W_1,\ldots,W_m$ in T such that $g(a_j) \in W_j$ for $1 \leqslant j \leqslant m$ and $W_1\cdots W_m \subset W$. Since g is continuous, there exist open sets $U_1, \ldots, U_m$ in X such that $a_j \in U_j$ and $g(U_j) \subset W_j$ for $1 \leqslant j \leqslant m$. Let $U = U_1 \times \cdots \times U_m$. Then U is open in F, $(a_1, \ldots, a_m) \in U$, and $f(U) \subset W$. It follows that f is continuous, and unique, since $F = \theta(\phi(X))$. The conclusion now follows from 2.59. ∎

<u>*2.61 Corollary*</u>. Let X be a Hausdorff space and (ϕ, F) the free topological semigroup generated by X. Then $\phi : X \to F$ is a homomorphism of X onto a subspace of F.

<u>*2.62 Corollary*</u>. Let X be a σ-compact Hausdorff space and (ϕ, F) the free topological semigroup generated by X. Then F is σ-compact.

Proof. Let $(X_n : n \in \mathbb{N}\}$ be a collection of compact spaces whose union is X. Then the compact sets of the form $X_{m_1} \times \cdots \times X_{m_n}$, for $m_j \in \mathbb{N}$ $(1 \leqslant j \leqslant n)$, $n \in \mathbb{N}$ constitute a countable collection whose union if F. ∎

<u>*2.63 Corollary*</u>. Let X be a locally compact Hausdorff space and (ϕ, F) the free topological semigroup generated by X. Then F is locally compact.

Proof. Let $(x_1, \ldots, x_n) \in F$ and $U_1 \times \cdots \times U_n$ be a basic open subset of F containing $(x_1, \ldots, x_n)$. Since X is locally compact, there exists an open set V_j containing x_j so that $\overline{V}_j$ is compact and $\overline{V}_j \subset U_j$ for $1 \leqslant j \leqslant n$. We obtain that

$$(x_1, \ldots, x_n) \in V_1 \times \cdots \times V_n \subset \overline{V}_1 \times \cdots \times \overline{V}_n \subset U_1 \times \cdots \times U_n,$$

and $\overline{V}_1 \times \cdots \times \overline{V}_n$ is compact. ∎

One can use a generalization of the technique employed in the proof of 2.57 to obtain free abelian topological semigroups, free topological groups, and free abelian topological groups. For this reason, we do not go into the details of showing the existence and uniqueness of these objects at this point. However, we will present a proof of the existence of free topological groups generated by locally compact, σ-compact Hausdorff spaces, since the technique used in this proof illustrates how a free topological group is obtained as a quotient of a free topological semigroup and yields additional information in the compact case.

A detailed study of the free topological group generated by a completely regular Hausdorff space appears in [Kakutani, 1944].

If X is a Hausdorff space, then a *free topological group generated by* X is a pair (γ,G) such that G is a topological group, $\gamma : X \to G$ is a continuous function, and if $g : X \to H$ is a continuous function from X into a topological group H, then there exists a unique continuous homomorphism $f : G \to H$ such that the diagram:

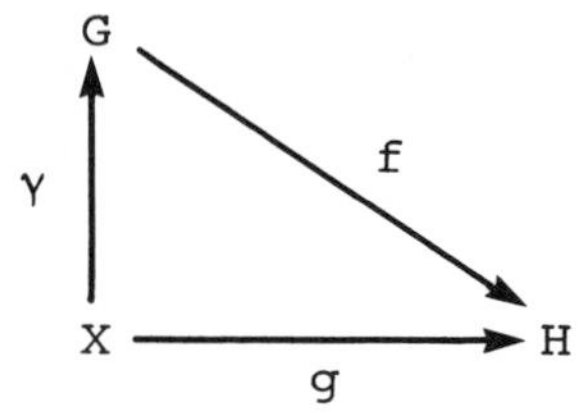

commutes.

2.64 Theorem. Let X be a locally compact σ-compact Hausdorff space. Then there exists a free topological group (γ,G) generated by X. Moreover, if X is compact, then $\gamma : X \to G$ is a homeomorphism of X onto a subspace of G.

Proof. Let Y be a space which is disjoint from X and homomorphic to X, and $h : X \to Y$ a homeomorphism. If $x \in X$, then we use x' to denote h(x), and if $x \in Y$, then we use x' to denote $h^{-1}(x)$. Let (S,ϕ) be the free topological semigroup generated by $X \cup Y$, and observe that S is locally compact and σ-compact from 2.62 to 2.63. It follows that S^1 is locally compact and σ-compact. (Recall that 1 is adjoined discretely.) Let $R_0 = \{(xx',1) : x \in X \cup Y\}$ and R the closed congruence on S^1 generated by R^0. Then, by 1.56, we have that

S^1/R is a topological semigroup. It is easily verified that S^1/R is algebraically a group. We will show that inversion is continuous on S^1/R. Let $\pi : S^1 \to S^1/R$ denote the natural map and $f : S^1 \to S^1$ the function defined by $(x_1,\ldots,x_n) \longmapsto (x_n',\ldots,x_1')$ and $1 \to 1$. Then f is continuous and the diagram:

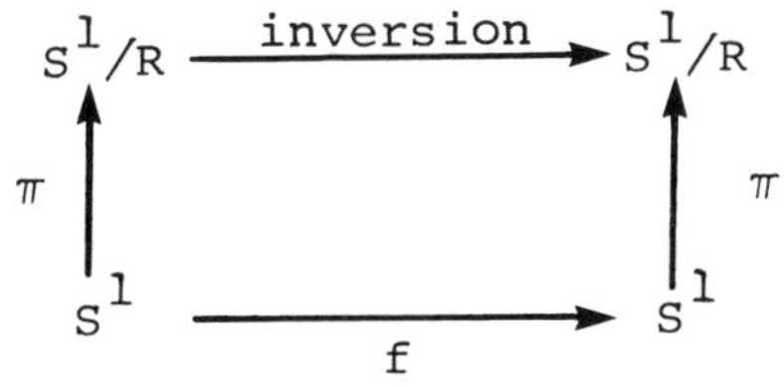

commutes. Since $\pi \circ f$ is continuous, and π is quotient, we obtain that inversion is continuous from 1.47, and hence S^1/R is a topological group. Let $G = S^1/R$ and $\gamma = \pi \circ j \circ \phi \circ i$, where $i : X \to X \cup Y$ and $j : S \to S^1$ are inclusions. It is now straightforward to show that (γ,G) is a free topological group generated by X, and the final conclusion is easily verified by proving that γ is injective. ∎

COPRODUCTS

In this section we present a brief discussion of coproducts of topological semigroups. It will be apparent from the definition that the concept of a coproduct is dual to that of product.

One of the major difficulties in dealing with coproducts of topological semigroups is that no general technique is available for determining their structure. For example, the structure of the coproduct of I_u and $\mathbb{R}/\mathbb{Z}$ is unknown.

If $\{S_\alpha : \alpha \in A\}$ is a collection of topological semigroups, then a *coproduct* of $\{S_\alpha : \alpha \in A\}$ is a collection $\{\phi_\alpha : S_\alpha \to S : \alpha \in A\}$ such that S is a topological semigroup, each ϕ_α is a continuous homomorphism, $S = \Gamma(\cup \{\phi_\alpha(S_\alpha) : \alpha \in A\})$, and if $\{\gamma_\alpha : S_\alpha \to T : \alpha \in A\}$ is a collection of continuous homomorphisms into a topological semigroup T, then there exists a continuous homomorphism $\phi : S \to T$ such that the diagram:

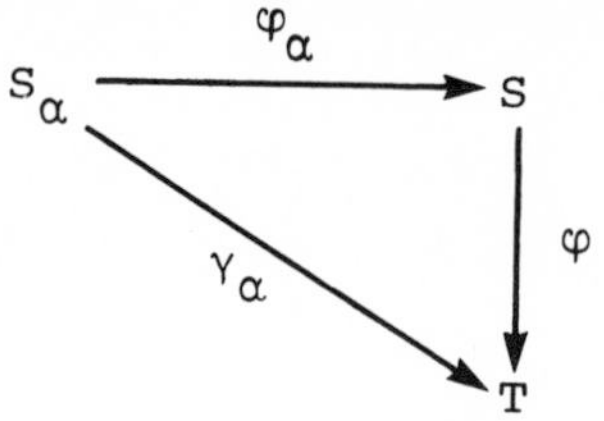

commutes for each $\alpha \in A$.

2.65 Theorem. (Uniqueness of Coproducts). Let $\{S_\alpha : \alpha \in A\}$ be a collection of topological semigroups. If $\{\phi_\alpha : S_\alpha \to M : \alpha \in A\}$ and $\{\mu_\alpha : S_\alpha \to N : \alpha \in A\}$ are coproducts of $\{S_\alpha : \alpha \in A\}$, then there exists a topological isomorphism $\sigma : M \to N$ such that the diagram:

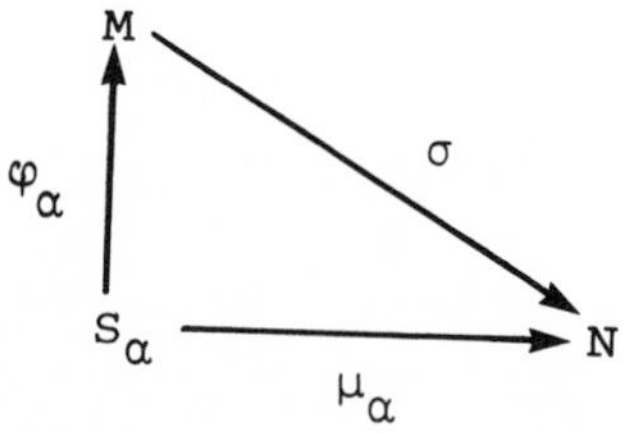

commutes for each $\alpha \in A$.

Proof. Since M is a coproduct, there is a continuous homomorphism σ such that the above diagram commutes for each $\alpha \in A$. On the other hand, using the fact that N is a coproduct, we have a continuous homomorphism $\phi : N \to M$ such that the diagram:

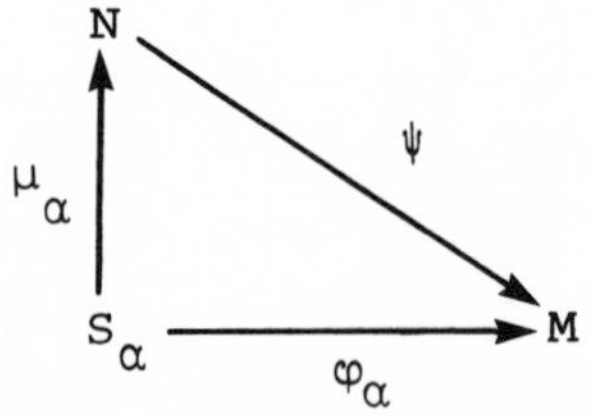

commutes for each $\alpha \in A$. Thus, for each $\alpha \in A$, we have $\phi_\alpha = \psi\mu_\alpha = \psi\sigma\phi_\alpha$, and hence $\psi\sigma|\phi_\alpha(S_\alpha) = 1_{\phi_\alpha(S_\alpha)}$. Let $Q = \cup\ \{\phi_\alpha(S_\alpha) : \alpha \in A\}$.

Then $\psi\sigma|Q = 1_Q$. Now for $x \in \theta(Q)$, we have $x = x_1x_2\cdots x_n$, where $x_j \in Q$, and hence $\psi\sigma(x) = \psi\sigma(x_1x_2\cdots x_n) = \psi\sigma(x_1)\cdots\psi\sigma(x_n) = x_1\cdots x_n = x$. We conclude that $\psi\sigma|\theta(Q)$ is dense in M, $\phi\sigma = 1_M$, and similarly $\sigma\psi = 1_N$. It follows that σ is a topological isomorphism. ■

2.66 Theorem. (Existence of Coproducts). Let $\{S_\alpha : \alpha \in A\}$ be a collection of topological semigroups. Then there exists a coproduct $\{\phi_\alpha : S_\alpha \to S : \alpha \in A\}$.

Proof. Let G denote the disjoint union of the collection $\{S_\alpha : \alpha \in A\}$, and $\Lambda = \{(T,\tau,m,\{\sigma_\alpha : \alpha \in A\}) : T \subset 2^{2^{2^G}}$, τ is a Hausdorff topology on T, m is a continuous associative multiplication on (T,τ), $\sigma_\alpha : S_\alpha \to T$ is a continuous homomorphism for each $\alpha \in A$, and $T = \Gamma(\cup\{\sigma_\alpha(S_\alpha) : \alpha \in A\})$. Then Λ is non-empty, since each singleton of $2^{2^{2^G}}$ admits the conditions required to be a member of Λ. A straightforward cardinality argument yields that Λ is a set. Let $P = \Pi\{(T,\tau,m) : (T,\tau,m,\{\sigma_\alpha : \alpha \in A\}) \in \Lambda\}$. Then P is a topological semigroup. Define $\phi_\alpha : S_\alpha \to P$ so that $\pi_{(T,\tau,m,\{\sigma_\alpha : \alpha \in A\})} \circ \phi_\alpha = \sigma_\alpha$ for each $\alpha \in A$. Then each ψ_α is a continuous homomorphism. Let $S = \Gamma(\cup\{\psi_\alpha(S_\alpha) : \alpha \in A\})$ and define $\phi_\alpha : S_\alpha \to S$ by $\phi_\alpha(x) = \psi_a(x)$ for each $x \in S_\alpha$ and each $\alpha \in A$.

We claim that $\{\phi_\alpha : S_\alpha \to S : \alpha \in A\}$ is a coproduct of $\{S_\alpha : \alpha \in A$. To establish this claim, suppose that M is a topological semigroup and $\{\gamma_\alpha : S_\alpha \to M : \alpha \in A\}$ is a collection of continuous homomorphisms. Let $S' = \Gamma(\cup \{\gamma_\alpha(S_\alpha) : \alpha \in A\})$ in M. In view of 2.42 we have, $\text{Card } S' \leqslant \text{Card } 2^{2^{\theta(\cup \{\gamma_\alpha(S_\alpha) : \alpha \in A\})}} \leqslant \text{Card } 2^{2^{2^{\cup \{\gamma_\alpha(S_\alpha) : \alpha \in A\}}}} \leqslant \text{Card } 2^{2^{2^G}}$. It follows that there exists a $T \subset 2^{2^{2^G}}$ and a bijection $k : S' \to T$. Let τ be the quotient topology on T induced by k, and define multiplication m on T to be the composition:

$$T \times T \xrightarrow{k^{-1} \times k^{-1}} S' \times S' \xrightarrow{m'} S' \xrightarrow{k} T ,$$

where m' is multiplication on S'. Then (T,τ,m) is a topological semigroup and $k : S' \to T$ is a topological isomorphism. Define $\sigma_\alpha : S_\alpha \to T$ by $\sigma_\alpha = k\gamma_\alpha$ for each $\alpha \in A$. Then $(T,\tau,m,\{\sigma_\alpha : \alpha \in A\}) \in \Lambda$. Define $\phi : S \to M$ by $\phi = k^{-1} \circ \pi_{(T,\tau,m,\{\sigma_\alpha : \alpha \in A\})}|S$. Then ϕ is a continuous homomorphism and the diagram:

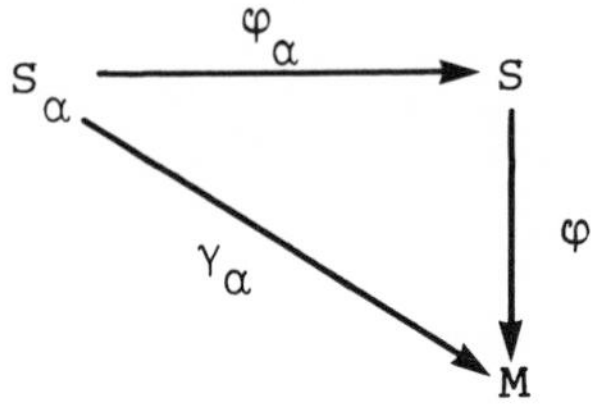

commutes for each $\alpha \in A$. It follows that $\{\phi_\alpha : S_\alpha \to S : \alpha \in A\}$ is the desired coproduct. ■

Slight modifications in the proofs of 2.65 and 2.66 yield the existence of unique coproducts in the category of compact semigroups and continuous homomorphisms, and in the category of compact monoids with identity preserving continuous homomorphisms. Some open questions in the theory of compact semigroups can be restated in terms of questions about coproducts in the latter category [Hofmann and Mostert, 1969].

Chapter 3

INTERNAL STRUCTURE THEOREMS

In this chapter we investigate the structure of some of the fundamental building blocks of a topological semigroup. We discuss monothetic semigroups, completely simple semigroups, and conclude the chapter by considering the various Green's relations and how they relate to the structure of semigroups.

MONOTHETIC SEMIGROUPS

A monothetic semigroup is the simplest building block of a topological semigroup, since it is the closed subsemigroup generated by a single element. We will investigate the structure of compact monothetic semigroups, construct the universal compact monothetic semigroup, and then turn briefly to the study of locally compact monothetic semigroups.

A semigroup S is said to be *cyclic* if there exists a surmorphism $\phi : \mathbb{N} \to S$. The element $\phi(1)$ is called a *(cyclic) generator* of S.

3.1 Theorem. Let S be a semigroup and $\phi : \mathbb{N} \to S$ a homomorphism. If $m, n \in \mathbb{N}$, $m < n$, and $\phi(n) = \phi(m)$ then $\phi(\mathbb{N}) = \phi(\{p \in \mathbb{N} : p \leqslant n\})$. Hence, either ϕ is injective or $\phi(\mathbb{N})$ is finite.

Proof. We will first prove by induction that $\phi(m + k(n - m)) = \phi(m)$ for each $k \in \mathbb{N}$. Now $\phi(m + 1 \cdot (n - m)) = \phi(n) = \phi(m)$. Assuming that $k \geqslant 2$ and that $\phi(m + (k - 1)(n - m)) = \phi(m)$, we have $\phi(m + k(n - m)) = \phi(m + (k - 1)(n - m) + (n - m)) = \phi(m + (k - 1)(n - m))\phi(n - m) = \phi(m)\phi(n - m) = \phi(m + n - m) = \phi(n) = \phi(m)$.

We claim that if $s \in \mathbb{N}$ and $m < s$, then $\phi(s + k(n - m)) = \phi(s)$ for each $k \in \mathbb{N}$. Now $\phi(s + k(n - m)) = \phi(s - m + m + k(n - m)) = \phi(s - m)\phi(m + k(n - m)) = \phi(s - m)\phi(m) = \phi(s - m + m) = \phi(s)$, in view of the preceding paragraph.

Finally, let $q \in \mathbb{N}$ with $n < q$. Then $q = m + k(n - m) + r$ for some $k \in \mathbb{N}$ and either $r = 0$ or $r \in \mathbb{N}$ with $r < n - m$. Now if $r = 0$, then $\phi(q) = \phi(m + k(n - m)) = \phi(m)$, by the first paragraph of this proof. Otherwise, $\phi(q) = \phi(m + k(n - m) + r) = \phi(m + r + k(n - m)) = \phi(m + r)$, since $m < m + r$. In view of the fact that $m + r < n$, the proof is complete. ■

A topological semigroup S is said to be *monothetic* if S contains a dense cyclic subsemigroup, i.e., there exists a dense homomorphism $\phi : \mathbb{N} \to S$. The element $\phi(1)$ is called a *(monothetic) generator* of S.

Recall, from Chapter 1, that if S is a topological semigroup and $x \in S$, then $\theta(x) = \{x^n : n \in \mathbb{N}\}$ is a subsemigroup of S. Clearly, the map $x \to x^n$ is a surmorphism of $\mathbb{N}$ onto $\theta(x)$, so that $\theta(x)$ is a cyclic subsemigroup of S with x as a generator. It is likewise clear that $\Gamma(x) = \overline{\theta(x)}$ is a monothetic subsemigroup of S with x as a generator. Note that if S is [locally] compact, then $\Gamma(x)$ is [locally] compact. Now $\Gamma(x)$ is abelian, and hence if $\Gamma(x)$ has a minimal ideal $M(\Gamma(x))$, then $M(\Gamma(x))$ is a group, e.g., if $\Gamma(x)$ is compact.

If a is an element of a topological semigroup S and $n \in \mathbb{N}$, then $\theta_n(a) = \{a^k : k \in \mathbb{N} \text{ and } n \leqslant k\}$. Observe that $\theta_1(a) = \theta(a)$.

The following basic result is due independently to [Numakura, 1952], [Peck, 1951], [Koch, 1957], and [Koch, 1953].

3.2 Theorem. Let $S = \Gamma(a)$ be a compact monothetic semigroup. Then the cluster points of the net $\{a^n\}$ form a group $M(a)$, and $M(a)$ is the minimal ideal of S. Moreover, if S has an identity, then S is either a finite group or is dense in itself.

Proof. Let C be the set of cluster points of $\{a^n\}$. We will show that $tC = C = Ct$ for each $t \in S$. Let $n \in \mathbb{N}$. Then

$$a^n C = a^n \cdot \overline{\{\theta_k(a) : k \in \mathbb{N}\}} = \cap \overline{\{a^n\,{}_k(a) : k \in \mathbb{N}\}}$$

$$= \cap \overline{\{\theta_{k+n}(a) : k \in \mathbb{N}\}} = \cap \overline{\{\theta_k(a) : k \in \mathbb{N}\}} = C$$

The second and third equality are consequences of the fact that left multiplication by a^n is a closed map. Thus, for each $n \in \mathbb{N}$, $a^n \in \{t \in S : tC = C\}$, which is closed. It follows that $tC = C$ for each $t \in S$ and, since S is abelian, $C = Ct$ for each $t \in S$. We conclude that $C = M(a)$ is a group.

If S has an identity 1, then for $p \neq q$ in S and $a^p = a^q$, we conclude from the first part that S is a group. Otherwise, if some $x \in S$ is isolated, then $x = a^p$ for some $p \in \mathbb{N}$. Since $\{a^n\}$ clusters to 1, $\{a^{p+n}\}$ clusters to $a^p = x$, and the result follows. ∎

3.3 Corollary. Let $S = \Gamma(a)$ be a compact monothetic semigroup. If a is not isolated, then S is a topological group.

3.4 Corollary. Let S be a compact monothetic semigroup with two distinct generators. Then S is a topological group.

Proof. Let a and b be distinct generators of S. The proof is divided into two cases, according as $b \in \theta(a)$ or $b \notin \theta(a)$. Suppose $b \in \theta(a)$. If $a \in \theta(b)$, then there are $m, n \in \mathbb{N}$ with $m, n > 1$ such that $b \in a^n$ and $a = b^m$. Hence $a \in a^{mn}$ where $mn > 1$, and it follows from 3.2 that S is a group. If $a \notin \theta(b)$, then a is not isolated and the result follows from 3.3. In the case $b \in \theta(a)$, then b is not isolated, and again the result follows from 3.3. ∎

The following is a reformulation of the first part of 3.2:

3.5 Theorem. Let $\Gamma(a)$ be a compact monothetic semigroup. Then $M(\Gamma(a))$ is a compact monothetic group with generator ea, where e is the identity of $M(\Gamma(a))$. Moreover, $\Gamma(a) = \theta(a) \cup M(\Gamma(a))$.

Observe in 3.5 that $\lambda_e : \Gamma(a) \to M(\Gamma(a))$ is a homomorphic retraction of $\Gamma(a)$ onto $M(\Gamma(a))$.

3.6 Theorem. Let G be a compact group. Then G is monothetic if and only if there exists a dense homomorphism $\psi : \mathbb{Z} \to G$.

Proof. Suppose that G is monothetic and let $\alpha : \mathbb{N} \to G$ be a dense homomorphism. Define $\psi(0) = e$ (the identity of G) and $\psi(-n) = \alpha(n)^{-1}$ (inverse in G) for each $n \in \mathbb{N}$, and $\phi(n) = \alpha(n)$ for each $n \in \mathbb{N}$. Then $\psi : \mathbb{Z} \to G$ is a dense homomorphism.

Suppose that $\psi : \mathbb{Z} \to G$ is given to be a dense homomorphism and define $\phi : \mathbb{N} \to G$ by $\phi(n) = \psi(n)$ for each $n \in \mathbb{N}$. Then ϕ is a homomorphism. To see that ϕ is dense, observe that $\overline{\phi(\mathbb{N})}$ is a compact subsemigroup of G and hence is a compact subgroup. Thus for $n \in \mathbb{N}$, $\phi(n)^{-1} \in \overline{\phi(\mathbb{N})}$ and $\phi(n)^{-1} = \psi(n)^{-1} = \psi(-n)$, and $\psi(0) = e \in \overline{\phi(\mathbb{N})}$. We obtain that $\psi(\mathbb{Z}) \subset \overline{\phi(\mathbb{N})}$, and hence $G = \overline{\psi(\mathbb{Z})} \subset \overline{\phi(\mathbb{N})}$, so that ϕ is dense. ■

Let $\mathbb{N}^* = \mathbb{N} \cup \{\infty\}$ denote the one point compactification of $\mathbb{N}$ with $\infty + x = x + \infty = \infty$ for all $x \in \mathbb{N}^*$, and let $b_{\mathbb{Z}} : \mathbb{Z} \to B(\mathbb{Z})$ denote the Bohr compactification of $\mathbb{Z}$. Observe that $\mathbb{N}^*$ is a compact monothetic semigroup.

We turn to the construction of the universal compact monothetic semigroup, which was first constructed in [Hofmann and Mostert, 1966]. It is, of course, the Bohr compactification of $\mathbb{N}$.

3.7 Theorem. Let $\mathbb{M} = \{(n,b_{\mathbb{Z}}(n)) : n \in \mathbb{N}\} \cup (\{\infty\} \times B(\mathbb{Z}))$ as a compact subsemigroup of $\mathbb{N}^* \times B(\mathbb{Z})$ and define $\beta : \mathbb{N} \to \mathbb{M}$ by $\beta(n) = (n,b_{\mathbb{Z}}(n))$ for each $n \in \mathbb{N}$. Then $\beta : \mathbb{N} \to \mathbb{M}$ is the Bohr compactification of $\mathbb{N}$.

Proof. Let S be a compact monothetic semigroup, $\phi : \mathbb{N} \to S$ a dense homomorphism, and let e denote the identity of M(S). Define $\psi : \mathbb{Z} \to M(S)$ by

$$\psi(n) = \begin{cases} \lambda_e \circ \phi(n) & \text{if } n < 0 \\ e & \text{if } n = 0 \\ (\lambda_e \circ \phi(-n))^{-1} & \text{if } n < 0 \end{cases}$$

Then ψ is a dense homomorphism, and the diagram:

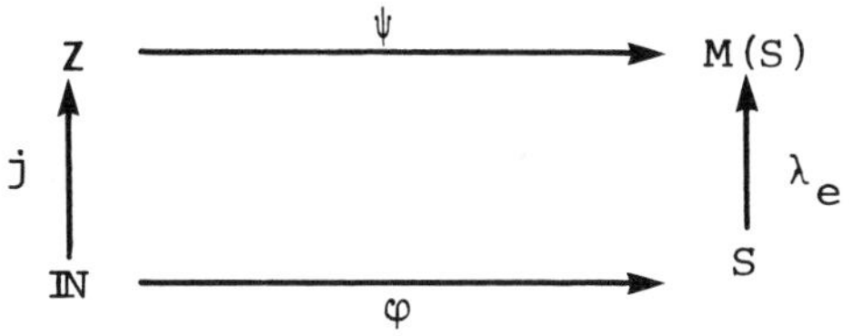

commutes, where $j : \mathbb{N} \to \mathbb{Z}$ is inclusion. We obtain a continuous surmorphism $\psi : B(\mathbb{Z}) \to M(S)$ such that the diagram:

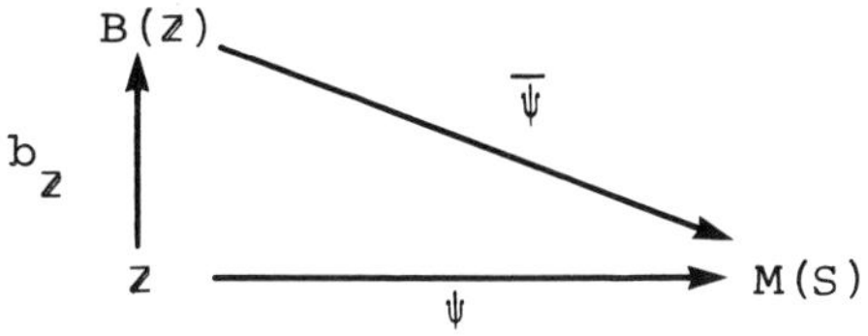

commutes. Define $\alpha : \mathbb{M} \to S$ by $\alpha(n, b_{\mathbb{Z}}(n)) = \phi(n)$ for $n \in \mathbb{N}$ and $\alpha(\infty, z) = \overline{\psi}(z)$ for $z \in B(\mathbb{Z})$. Then α is a continuous homomorphism such that the diagram:

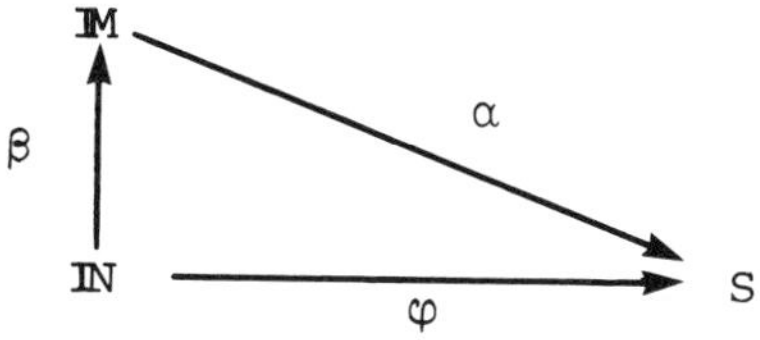

commutes. It is clear that $\beta(\mathbb{N})$ is dense in $\mathbb{M}$, and we conclude that $\beta : \mathbb{N} \to \mathbb{M}$ is the Bohr compactification of $\mathbb{N}$. ■

3.8 Corollary. A compact semigroup S is monothetic if and only if there exists a continuous surmorphism ϕ : $\mathbb{M} \to S$.

The semigroup $\mathbb{M}$, in view of 3.7 and 3.8, is called the *universal compact monothetic semigroup.*

As in [Hofmann and Mostert, 1966] we will characterize the surmorphic images of $\mathbb{M}$ by establishing that each continuous surmorphism on $\mathbb{M}$ can be lifted to $\mathbb{N}^* \times B(\mathbb{Z})$. We will employ the following:

3.9 Lemma. Let S be a compact abelian semigroup, e the identity of the compact group M(S), and $\pi : S \to S/M(S)$ the natural map. Define $\pi_S : S \to S/M(S) \times M(S)$ by $\pi_S(x) = (\pi(x), \lambda_e(x))$. Then π_S is an embedding.

Proof. This is straightforward. ∎

We will adopt π_S for the embedding in 3.9.

3.10 Theorem. Let S be a compact monothetic semigroup and $\alpha : \mathbb{M} \to S$ a continuous surmorphism. Then there exists a continuous surmorphism γ such that the diagram:

$$\begin{array}{ccc} \mathbb{N}^* \times B(\mathbb{Z}) & \xrightarrow{\ \gamma\ } & S/M(S) \times M(S) \\ {\scriptstyle j}\uparrow & & \uparrow{\scriptstyle \pi_S} \\ \mathbb{M} & \xrightarrow[\ \alpha\ ]{} & S \end{array}$$

commutes, where $j : \mathbb{M} \to \mathbb{N}^* \times B(\mathbb{Z})$ is inclusion.

Proof. We will refer to the maps in the proof of 3.7. Define $\phi : \mathbb{N} \to S$ by $\phi = \alpha \circ \beta$, where $\beta : \mathbb{N} \to \mathbb{M}$ is the Bohr compactification of $\mathbb{N}$. Obtain the continuous surmorphism $\overline{\psi} : B(\mathbb{Z}) \to M(S)$ as in 3.7. Define $\delta : \mathbb{N}^* \to S/M(S)$ by

$$\delta(n) = \Pi \circ \begin{cases} \alpha(n, b_{\mathbb{Z}}(n)) & \text{if } n \in \mathbb{N} \text{ and} \\ \alpha(\infty, f) & \text{if } n = \infty \end{cases}$$

where f is the identity of $B(\mathbb{Z})$, and $\pi : S \to M(S)$ is the natural map. Then δ is a continuous surmorphism, and $\gamma = \delta \times \overline{\psi}$ is the desired map. ∎

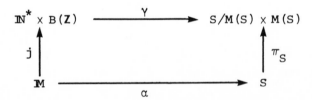

To complete the characterization of 3.10, we observe, in view of 3.5, that if S is a compact monothetic semigroup, then $S/M(S) \overset{I}{\approx} \mathbb{N}^*$ if S/M(S) is infinite, and $S/M(S) \overset{I}{\approx} \mathbb{N}^*/J(n)$ for some $n \in \mathbb{N}$ if S/M(S) is finite.

We turn now to locally compact monothetic semigroups. In this situation much less is known about the structure than in the case of compact monothetic semigroups. We begin with a classical result due to A. Weil. The formulation is slightly modified for use in a later theorem. See [Weil, 1938].

3.11 Theorem. Let G be a locally compact group and $\phi : \mathbb{Z} \to G$ a dense continuous homomorphism. Then either G is compact or ϕ is a topological isomorphism of $\mathbb{Z}$ onto G.

3.12 Lemma. Let S be a locally compact semigroup such that M(S) is non-empty and compact. Then for each open set V containing M(S), there exists an open set J such that $M(S) \subset J \subset V$ and J is an ideal of $\overline{J}$.

Proof. Let V be an open subset of S containing M(S). Since S is locally compact and M(S) is compact, we can assume that $\overline{V}$ is compact. Observe that $M(S)\overline{V} = M(S) \subset V$, and hence by 1.1, there exists an open set W such that $M(S) \subset W \subset V$ and $W\overline{V} \subset V$. Since $W \subset V$, we have $W^2 \subset V$ and inductively $\cup \{W^n : n \in \mathbb{N}\} \subset V$. Let $P = \cup \{W^n : n \in \mathbb{N}\}$. Then, since $\overline{V}$ is compact, P is a compact subsemigroup of S containing M(S). Let J be the union of all ideals of P contained in W. Then J is a non-empty ideal, since $M(S) \subset W$. We claim that J is open. Let $x \in J$. Then $x \cup Px \cup xP \cup PxP \subset J \subset W$. Since P is compact and W is open, again using 1.1, there exists an open set N containing x such that $N \cup PN \cup NP \cup PNP \subset W$. Since $N \cup PN \cup NP \cup PNP$ is an ideal of P contained in W it lies in J. Hence, $x \in N \subset J$, and J is open. Since J is an ideal of P, we have that J is an ideal of $\overline{J}$, and the proof of the lemma is complete. ∎

The following result appears in [Koch, 1957].

3.13 Theorem. Let S be a locally compact monothetic semigroup with a minimal ideal. Then S is compact.

Proof. Now M(S) is a group and $\lambda_e : S \to M(S)$ is a retraction of S onto M(S), where e is the identity of M(S). It follows that M(S) is locally compact, and hence a topological group. We observe that there exists a dense homomorphism $\phi : \mathbb{N} \to M(S)$, using the fact that S is monothetic and $\lambda_e : S \to M(S)$ is a homomorphic retraction. Since M(S) is a topological group, we obtain a dense homomorphism $\psi : Z \to M(S)$. In view of the local compactness and 3.11, we have either M(S) is compact or $M(S) \overset{T}{\approx} \mathbb{Z}$. Since ϕ is dense, we must conclude that M(S) is compact. By 3.12, there exists an open subsemigroup J of S containing M(S) so that $\overline{J}$ is compact. Now let a be a monothetic generator of S. In view of 3.1, there exists $n \in \mathbb{N}$ such that $a^n \in J$, and hence $\Gamma(a^n) \subset \overline{J}$ and $\Gamma(a^n)$ is compact. Let $T = \{a, a^2, \ldots, a^{n-1}\}$. Then $S = \Gamma(a) = T \cup T \cdot \Gamma(a^n)$, which is compact. This completes the proof. ■

If S is a compact monothetic semigroup with $\sigma : \mathbb{N} \to S$ a dense homomorphism, then according to 3.1, either σ is injective or $\sigma(\mathbb{N})$ is infinite. In the case that σ is injective, the typical structure of S is like that of $\mathbb{M}$ with the minimal ideal replaced by a given compact monothetic group. Such a semigroup can be constructed in the same manner as the construction of $\mathbb{M}$ in 3.7. On the other hand, if $\sigma : \mathbb{N} \to S$ is a surmorphism of $\mathbb{N}$ onto a nondegenerate finite semigroup S, then $S = \{\sigma(1), \ldots, \sigma(m - 1)\}$ and $M(S) = \sigma\{(n), \ldots, \sigma(m - 1)$ is the finite cyclic group of order m - n, where $n = \min \{p \in \mathbb{N} : \sigma(p) = \sigma(q), q \in \mathbb{N}, q \neq p\}$ and $m = \min \{p \in \mathbb{N} : n < p \text{ and } \sigma(p) = \sigma(n)\}$.

COMPLETELY SIMPLE SEMIGROUPS

Here we determine the structure of simple semigroups which contain minimal left and right ideals, as well as topological results which relate a compact semigroup to its minimal ideal. The algebraic structure is due to Sushkewitsch, Rees, and Clifford. The topological results are due to Wallace.

A semigroup S is said to be *simple* if for each ideal I of S, I = S, i.e., S contains no proper ideals. A simple semigroup S is said to be *completely simple* provided S contains a minimal left ideal and a minimal right ideal.

If S is a semigroup, then $M_L(S)$ and $M_R(S)$ denote the collection of all minimal left ideals and right ideals of S, respectively.

If no confusion seems likely, we write simple M_L and M_R for $M_L(S)$ and $M_R(S)$, respectively.

3.14 Lemma. Let S be a semigroup. Then S is simple if and only if SaS = S for each $a \in S$.

Proof. If S is simple, then the observation that SaS is an ideal for each $a \in S$ leads to the conclusion that S = SaS.

Suppose that S = SaS for each $a \in S$, and let I be an ideal of S. Let $a \in I$. Then $S = Sas \subset SIS \subset I$, and we conclude that S is simple. ∎

3.15 Lemma. Let S be a semigroup, L a minimal left ideal of S, and $a \in L$. Then La = Sa = L.

Proof. Note that $La \subset Sa \subset SL \subset L$. Since La is a left ideal of S and L is a minimal left ideal of S, we obtain that L = La, and the conclusion follows. ∎

3.16 Lemma. Let S be a semigroup, L a minimal left ideal of S, and $c \in S$. Then Lc is a minimal left ideal of S.

Proof. It is clear that Lc is a left ideal of S. Suppose that N is a left ideal of S such that $N \subset Lc$. Let $L_1 = \{x \in L : xc \in N\}$. Then L_1 is a left ideal of S such that $L_1 \subset L$. In view of the minimality of L, we see that $L_1 = L$, so that $Lc \subset N$, and the conclusion follows. ∎

3.17 Corollary. Let S be a semigroup with $M_L \neq \square \neq M_R$. Then $\cup M_L = \cup M_R = \cup \{R \cap L : R \in M_R, L \in M_L\}$ is the unique minimal ideal of S.

Proof. In view of 3.16 and its obvious dual, we see that $\cup M_L$ and $\cup M_R$ are ideals of S. Let T be an ideal of S. Then for each

$L \in M_L$, we have $\square \neq TL \subset T \cap L$, and hence $L \subset T$, since $T \cap L$ is a left ideal contained in L and L is minimal. Thus $\cup M_L \subset T$, and similarly $\cup M_R \subset T$. The result follows. ∎

3.18 Lemma. Let S be a semigroup such that $M_L \neq \square \neq M_R$. If $L \in M_L$ and $R \in M_R$, then $RL = R \cap L = eSe = H(e)$ for some $e \in E(S)$. Moreover, $R = eS$ and $L = Se$.

Proof. Note that RL is a semigroup. Let $x \in RL \subset R \cap L$. We will show that $xRL = RL = RLx$. Since $x \in R$, we conclude, from the dual of 2.4, that $xR = R$, so that $xRL = RL$. Similarly, $Lx = L$, $RLx = RL$, and hence $xRL = RL = RLx$. Thus RL is a group. Let e denote the identity of RL. Since $e \in R \cap L$, we have using 2.4 and its dual, that $R = eS$ and $L = Se$. Thus $R \cap L = eS \cap Se = eSe = eS(e) \subset eS(Se) = RL$. We conclude that eSe is a group, and $eSe = H(e)$. ∎

Recall, from Chapter 2, that if X and Y are nonempty sets, G is a group, and $\sigma : Y \times X \to G$ is a function, then $[X,G,Y]_\sigma$ is a semigroup on $X \times G \times Y$ with multiplication $((x,g,y), (x',g',y')) \to (x,g\sigma(y,x')g',y')$.

3.19 Lemma. Let X and Y be non-empty sets, G a group, and $\sigma : Y \times X \to G$ a function. Then

(a) $[X,G,Y]_\sigma$ is a simple semigroup;

(b) $M_L \neq \square$ and the minimal left ideals of $[X,G,Y]_\sigma$ are of the form $X \times G \times \{y\}$ for $y \in Y$;

(c) $M_R \neq \square$ and the minimal right ideals of $[X,G,Y]_\sigma$ are of the form $\{x\} \times G \times Y$ for $x \in X$; and

(d) the maximal subgroups of $[X,G,Y]_\sigma$ are of the form $\{x\} \times G \times \{y\}$ for $x \in X$ and $y \in Y$, and each is isomorphic to G.

Proof. To prove (a), let $(x,g,y) \in [X,G,Y]_\sigma$. Then $[X,G,Y]_\sigma(x,g,y)[X,G,Y]_\sigma = X \times G\sigma(Y \times \{x\})g\sigma(\{y\} \times X)G \times Y = X \times G \times Y$. Part (a) now following from 3.16.

To prove (b), first observe that $X \times G \times \{y\}$ is a left ideal. Let L be a left ideal of $[X,G,Y]_\sigma$ contained in $X \times G \times \{y\}$, and let

$(x,g,y) \in L$. Then $[X,G,Y]_\sigma(x,g,y) = X \times G \times \{y\} \subset L$, and (b) is proved.

The proof of (c) is similar to the proof of (b).

To prove (d), observe that $\{x\} \times G \times \{y\}$ is a maximal group from 3.18 using parts (a) and (b). Define $\alpha : G \to \{x\} \times G \times \{y\}$ by $\alpha(g) = (x, g\sigma(y,x)^{-1}, y)$. Note that $\alpha(g)\alpha(h) = (x, g\sigma(y,x)^{-1}\sigma(y,x) h\sigma(y,x)^{-1}, y = (x, gh\sigma(y,x)^{-1}, y) = \alpha(gh)$, so that α is a homomorphism. Define $\beta : \{x\} \times G \times \{y\} \to G$ by $\beta(x,g,y) = g\sigma(y,x)$. It is straightforward to show that α and β are mutually inverse homomorphisms, so that $\{x\} \times G \times \{y\}$ is isomorphic to G. ■

3.20 Theorem. Let S be a completely simple semigroup and $e \in E$. Then eSe is a group. Define $\sigma : (eS \cap E) \times (Se \cap E) \to eSe$ by $\sigma(h,f) = hf$, $\phi : [Se \cap E, eSe, eS \cap E]_\sigma \to S$ by $\phi(f,g,h) = fgh$, and $\hat{\phi} : S \to [Se \cap E, eSe, eS \cap E]_\sigma$ by $\hat{\phi}(s) = (s(eSe)^{-1}, ese, (ese)^{-1}s)$. Then ϕ and $\hat{\phi}$ are mutually inverse isomorphisms.

Conversely, if X and Y are non-empty sets, G is a group, and $\sigma : Y \times X \to G$ is a function, then $[X,G,Y]_\sigma$ is a completely simple semigroup.

Proof. The fact that eSe is a group follows readily from 3.17 and 3.18. We note that $\hat{\phi}$ is well-defined by showing, for example, that $s(ese)^{-1} \in Se \cap E$ for $s \in S$. In fact, $s(ese)^{-1}s(ese)^{-1} = s(ese)^{-1}(ese)(ese)^{-1} = s(ese)^{-1} = Se$. To see that ϕ is a homomorphism, note that $\phi[(r,g,\ell)(s,h,t)] = \phi(r, g\sigma(\ell,s)h, t) = rg\sigma(\ell,s)ht = rg\ell sht = \phi(r,g,\ell)\phi(s,h,t)$. Now $\phi\hat{\phi}(s) = \phi(s(ese)^{-1}, ese, (ese)^{-1}s) = s(ese)^{-1}(ese)(ese)^{-1}s = s(ese)^{-1}s$. Since $s(ese)^{-1} \in E \cap sS$ and sS is a minimal right ideal of S containing s, $s(ese)^{-1}S = sS$ and $s(ese)^{-1}s = s$ (3.19 and 3.17). Thus $\phi\hat{\phi}(f,g,h) = 1_S$. We next show that $\hat{\phi}\phi = 1_{[Se \cap E,\ eSe,\ eS \cap E]_\sigma}$. Indeed $\hat{\phi}\phi(f,g,h) = \hat{\phi}(fgh) = (fgh(efghe)^{-1}, efghe, (efghe)^{-1}fgh)$. Since $f \in Se \cap E$, we have $Sf = Se$, so that $f = fe$ and $e = ef$. Likewise, $h \in eS \cap E$, $LS = eS$, $h = eh$, and $e = Le$. Thus $fgh(efghe)^{-1} = fgh(ege)^{-1} = fghg^{-1} = fgheg^{-1} = fgg^{-1} = fe = f$. Similarly, $efghe = g$ and $(efghe)^{-1}fgh = h$, and so

$\hat{\phi}\phi(g,g,h) = (f,g,h)$. It follows that ϕ and ϕ are mutually inverse isomorphisms.

The converse is immediate from 3.19. ■

As an illustration of the structure theorem, we display an 8-element completely simple semigroup whose idempotents do not form a subsemigroup. Let $X = \{1,2\} = Y$, and $G = \{1,-1\}$ under multiplication. Define $\sigma : Y \times X \to G$ by $\sigma(1,1) = \sigma(2,1) = \sigma(2,2) = 1$ and $\sigma(1,2) = -1$. Then $[X,G,Y]_\sigma = \{(1,1,1),(1,1,2), (2,1,1), (2,1,2),(1,-1,1),(1,-1,2)$ $(2,-1,1),(2,-1,2)\}$. The idempotents are $(1,1,1)$, $(1,1,2)$, $(2,1,2)$, and $(2,-1,2)$. Note that $(1,1,1)(2,1,2) = (1,-1,2)$.

We now present a topological semigroup analog of 3.20.

3.21 Theorem. Let S be a topological semigroup with $M_L \neq \square \neq M_R$. Then M(S) is a completely simple semigroup. Suppose that M(S) is locally compact, and let $e \in M(S) \cap E(S)$. Then eSe is a locally compact group. Define: $\sigma : (eS \cap E(S)) \times (Se \cap E(S)) \to ESe$ by $\sigma(h,f) = hf$, $\phi : [Se \cap E(S), eSe, eS \cap E(S)]_\sigma \to M(S)$ by $\phi(f,g,h) = fgh$, and $\hat{\phi} : M(S) \to [Se \cap E(S), eSe, eS \cap E(S)]_\sigma$ by $\hat{\phi}(s) = (s(ese)^{-1}, ese, (ese)^{-1}s)$. Then ϕ and $\hat{\phi}$ are mutually inverse topological isomorphisms.

Conversely, if X and Y are locally compact Hausdorff spaces, G is a locally compact group, and $\sigma : Y \times X \to G$ is a completely simple locally compact semigroup.

Proof. Now eSe is closed in M(S), and hence is locally compact. In view of 3.20, we see that eSe is algebraically a group. The fact that eSe is a locally compact group now follows. Since multiplication in S is continuous, so is ϕ. The fact that $\hat{\phi}$ is continuous follows from the continuity of multiplication and the continuity of inversion in eSe. ■

Suppose that S is a completely simple locally compact semigroup. Note that $M_L[M_R]$ acquires in a natural way a right zero [left zero] multiplication, since if $L_1, L_2 \in M_L[R_1, R_2 \in M]$, then $L_1L_2 = L_2[R_1R_2 = R_1]$. We give M_L and M_R the quotient topology induced by the natural maps $\alpha_1 : S \to M_L$ and $\alpha_2 : S \to M_R$, respectively. Define $\hat{\alpha}_1 : M_R \to Se \cap E(S)$ and $\hat{\alpha}_2 : M_L \to eS \cap E(S)$, by $\hat{\alpha}_1(R) = E(R \cap Se)$ and $\hat{\alpha}_2(L) = E(eS \cap L)$, where $e \in E(S)$, $E(R \cap Se)$ is the idempotent in $R \cap Se$, and $E(eS \cap L)$ is the idempotent in $eS \cap L$. Define $\tau_1 : S \to Se \cap E(S)$ by $\tau_1(x) = x(exe)^{-1}$, and $\tau_2 : S \to eS \cap E(S)$ by $\tau_2(x) = (exe)^{-1}x$. The diagrams:

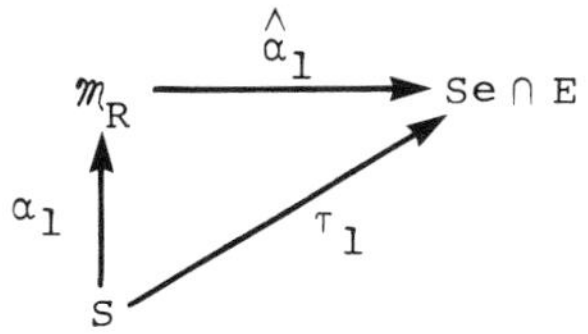

and

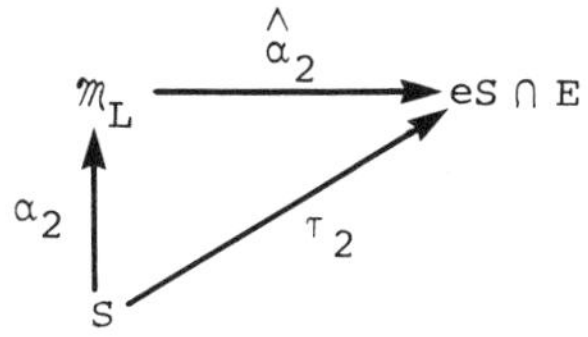

commute.

Using the notation in 3.21 we have:

3.22 Theorem. Let S be a completely simple locally compact semigroup. Then τ_1 and τ_2 are open maps, and $\hat{\alpha}_1 : M_R \to Se \cap E$ and $\hat{\alpha}_2 : M_L \to eS \cap E$ are topological isomorphisms.

Proof. Note that $\tau_1 = \pi_1\hat{\phi}$ and $\tau_2 = \pi_2\hat{\phi}$, where $\hat{\phi}$ is the topological isomorphism in 3.21 and π_1 and π_2 are the first and second projections of $[Se \cap E, eSe, eS \cap E]_\sigma$. It follows that τ_1 and τ_2 are open. Thus $\hat{\alpha}_1$ and $\hat{\alpha}_2$ are open, since α_1 and α_2 are quotient maps. Further, α_1 and α_2 are injective, since M_R and M_L are disjoint. Finally, $Se \cap E$ has left zero multiplication, since $f, g \in Se \cap E$ implies $Sf = Sg$, and $fg = f$, and dually for $eS \cap E$. ∎

We next identify various retracts of a topological semigroup with a locally compact minimal ideal.

3.23 Theorem. Let S be a topological semigroup with a locally compact minimal ideal M, and let $e \in M \cap E$. Then

(a) $x \to E(xe) = x(exe)^{-1}$ is a retraction of S onto $Se \cap E$;

(b) $x \to E(ex) = (exe)^{-1}x$ is a retraction of S onto $eS \cap E$;

(c) $x \to xE(ex) = x(exe)^{-1}x$ is a retraction of S onto M; and

(d) $x \to E(x) = xx^{-1}$ is a retraction of M onto $E \cap M$.

Proof. (a), (b), and (c) follow from the fact that eSe = eMe is a retract of M, and is hence closed and locally compact, together with the continuity of inversion and multiplication in eSe.

To prove (d), we first show that inversion is continuous in $[Se \cap E, ese, eS \cap E]_\sigma$ (see 3.21). Let $\{(x_\alpha, g_\alpha, y_\alpha)\}$ be a net in $[Se \cap E, eSe, eS \cap E]_\sigma$ converging to (x,g,y). Then $(x_\alpha, g_\alpha, y_\alpha)^{-1} = (x_\alpha, \sigma(y_\alpha, x_\alpha)^{-1} g_\alpha^{-1} \sigma(y_\alpha, x_\alpha)^{-1}, y_\alpha) \to (x, \sigma(y,x)^{-1} g^{-1} \sigma(x,y)^{-1}, y) = (x,g,y)^{-1}$, since multiplication and inversion are continuous in eSe. It follows that inversion is continuous in M, so that $x \to xx^{-1}$ is continuous. ∎

Recall that a point p of a connected space X is a *cutpoint* of X if $X \backslash \{p\}$ is not connected. We use the term *continuum* to mean a compact connected Hausdorff space. Recall also that each non-degenerate continuum has at least two non-cutpoints.

An interesting application of the structure theorem for completely simple topological semigroups is the following result of Faucett:

3.24 Theorem. Let S be a topological semigroup with a compact connected minimal ideal. If M(S) has a cutpoint, then M(S) consists of left zeros or of right zeros.

Proof. Let $e \in M \cap E$. Then, by 3.21 we have $M(S) \overset{T}{\approx} (Se \cap E) \times eSe \times (eS \cap E)$. Also, each factor is a retract of M(S) and hence is a continuum. Since the Cartesian product of two non-degenerate continua cannot contain a cutpoint, it follows that two of the three factors must be degenerate. If $Se \cap E$ and eSe are degenerate, then $M(S) \overset{T}{\approx} \{e\} \times \{e\} \times (eS \cap E) \overset{T}{\approx} M_L$, and thus M(S) has the right zero multiplication. A similar argument works to show that M(S) has left zero multiplication in case $eS \cap E$ and eSe are degenerate. In case $Se \cap E$ and $eS \cap E$ are degenerate, we conclude that M(S) is a group. Thus by homogeneity, each element of M(S) must be a cutpoint. It follows that the continuum M(S) is degenerate, so $M = \{0\}$. ∎

GREEN'S RELATION

Here we investigate the various Green's relation and how they reveal some insight into the structure of semigroups. Considerable portions of the material in this section are algebraic in nature and appear in [Green, 1951] and [Clifford and Preston, 1961]. The concept of stability in a semigroup was investigated in [Koch and Wallace, 1957] and explored further in [Anderson, Hunter, and Koch, 1965]. Topological results appearing in this section can be found in [Wallace, 1953] and [Hofmann and Mostert, 1966].

Recall, from Chapter 1, that if S is a semigroup, and $a \in S$, then $L(a) = S^1a$ (the left ideal of S generated by a), $R(a) = aS^1$ (the right ideal of S generated by a), and $J(a) = S^1aS^1$ (the ideal of S generated by a).

If S is a semigroup, then we define

(a) $\mathcal{L}(S) = \{(a,b) \in S \times S : L(a) = L(b)\}$;

(b) $\mathcal{R}(S) = \{(a,b) \in S \times S : R(a) = R(b)\}$;

(c) $\mathcal{J}(S) = \{(a,b) \in S \times S : J(a) = J(b)\}$;

(d) $\mathcal{H}(S) = \{(a,b) \in S \times S : L(a) = L(b) \text{ and } R(a) = R(b)\}$.

If no confusion seems likely, we write $\mathcal{L}$, $\mathcal{R}$, $\mathcal{J}$, and $\mathcal{H}$ in place of $\mathcal{L}(S)$, $\mathcal{R}(S)$, $\mathcal{J}(S)$, and $\mathcal{H}(S)$, respectively. Occasionally one writes $a\mathcal{L}b$ to mean $(a,b) \in \mathcal{L}$ and similarly for $\mathcal{R}$, $\mathcal{J}$, and $\mathcal{H}$.

3.25 Theorem. Let S be a semigroup. Then $\mathcal{L}$, $\mathcal{R}$, $\mathcal{J}$, and $\mathcal{H}$ are equivalence relations on S. Moreover, $\mathcal{L}$ is a right congruence on S, and $\mathcal{R}$ is a left congruence on S.

Proof. It is immediate from their definitions that $\mathcal{L}$, $\mathcal{R}$, and $\mathcal{J}$ are equivalence relations on S. That $\mathcal{H}$ is an equivalence relation follows from the observation that $\mathcal{H} = \mathcal{L} \cap \mathcal{R}$.

To see that $\mathcal{L}$ is a right congruence, let $(a,b) \in \mathcal{L}$ and $x \in S$. Then $S^1a = S^1b$, and hence $S^1ax = S^1bx$, or equivalently $(ax,bx) \in \mathcal{L}$. It follows that $\mathcal{L}$ is a right congruence on S. The proof that $\mathcal{R}$ is a left congruence is similar. ■

The next theorem will be used to define the $\mathcal{D}$ relation on a semigroup. Later in this section, we will prove that if S is either a compact or abelian semigroup, then $\mathcal{D} = \mathcal{J}$.

3.26 Theorem. If S is a semigroup, then $\mathcal{L} \circ \mathcal{R} = \mathcal{R} \circ \mathcal{L}$.

Proof. Let $(a,b) \in \mathcal{L} \circ \mathcal{R}$. Then $(a,x) \in \mathcal{L}$ and $(x,b) \in \mathcal{R}$ for some $x \in S$. It follows that $a = cx$ and $b = xd$ for some $c, d \in S^1$. Let $t = ad = cxd = cb$. Since $\mathcal{L}$ is a right congruence and $\mathcal{R}$ is a left congruence, we have $(t,b) = (ad,xd) \in \mathcal{L}$ and $(a,t) = (cx,cb) \in \mathcal{R}$, and hence $(a,b) \in \mathcal{R} \circ \mathcal{L}$. This proves that $\mathcal{L} \circ \mathcal{R} \subset \mathcal{R} \circ \mathcal{L}$. A similar argument works to show that $\mathcal{R} \circ \mathcal{L} \subset \mathcal{L} \circ \mathcal{R}$, and equality results. ∎

If S is a semigroup, then we define $\mathcal{D}(S) = \mathcal{R}(S) \circ \mathcal{L}(S) = \mathcal{L}(S) \circ \mathcal{L}(S) \circ \mathcal{R}(S)$.

As with the other Green's relations, when no confusion seems likely, we write $\mathcal{D}$ in place of $\mathcal{D}(S)$, and occasionally write $a\mathcal{D}b$ to mean $(a,b) \in \mathcal{D}$.

The following is a consequence of 3.26 and the definition of $\mathcal{D}$:

3.27 Theorem. If S is a semigroup, then $\mathcal{D}$ is an equivalence relation on S.

If S is a semigroup and $a \in S$, then $L_a(S)$, $R_a(S)$, $J_a(S)$, $H_a(S)$, and $D_a(S)$ denote the $\mathcal{L}$, $\mathcal{R}$, $\mathcal{J}$, $\mathcal{H}$, and $\mathcal{D}$ classes of a in S, respectively. Again, if no confusion is likely, we write simply L_a, R_a, J_a, H_a, and D_a.

The next theorem is an important connecting theorem for Green's relations on a semigroup.

3.28 Theorem. If S is a semigroup, then $\mathcal{H} = \mathcal{L} \cap \mathcal{R} \subset \mathcal{L} \cup \mathcal{L} \cup \mathcal{R} \subset \mathcal{D} \subset \mathcal{J}$.

Proof. It is clear that $\mathcal{H} = \mathcal{L} \cap \mathcal{R} \subset \mathcal{L} \cup \mathcal{R} \subset \mathcal{D}$.

To see that $\mathcal{D} \subset \mathcal{J}$, let $(a,b) \in \mathcal{D}$. Then $(a,x) \in \mathcal{L}$ and $(x,b) \in \mathcal{R}$ for some $x \in S$. We obtain that $S^1a = S^1x$ and $xS^1 = bS^1$. It follows that $S^1aS^1 = S^1xS^1 = S^1bS^1$, and hence $(a,b) \in \mathcal{J}$. We conclude that $\mathcal{D} \subset \mathcal{J}$. ∎

3.29 Corollary. Let S be an abelian semigroup. Then $\mathcal{L} = \mathcal{R} = \mathcal{J} = \mathcal{H} = \mathcal{D}$, and $\mathcal{D}$ is a congruence on S.

Proof. Since S is abelian, it is clear that $\mathcal{L} = \mathcal{R} = \mathcal{J}$. The conclusion now follows from 3.28 and the fact that $\mathcal{L}$ is a right congruence and $\mathcal{R}$ is a left congruence. ■

We now present a proof of the result that the $\mathcal{H}$-class of an idempotent is a group.

3.30 Theorem. Let S be a semigroup and $e \in E(S)$. Then $H_e = H(e)$.

Proof. Now $H(e) = eSe \cap \{x \in S : e \in xS \cap Sx\}$.

Let $a \in H_e$. Then $e \in Sa$, and hence $L(e) \subset L(a)$. We also have that $a \in eSe \subset Se$, so that $L(a) \subset L(e)$. This proves that $L(a) = L(e)$, and similarly, $R(a) = R(e)$. We obtain that $H(e) \subset H_e$.

Let $b \in H_e$. Then $R(b) = R(e)$ and $L(b) = L(e)$. We can assume that $b \neq e$. From $e \in R(b) = bS^1$ and $e \in L(b) = S^1b$, we obtain that $e \in bS \cap Sb$. Now $b \in eS$ and $b \in Se$ yields that $b \in eSe$, and hence $b \in H(e)$. This proves that $H_e \subset H(e)$, and we conclude that $H_e = H(e)$. ■

We now investigate the notion of stability of a semigroup and employ this concept to show that $\mathcal{D} = \mathcal{J}$ for a compact semigroup.

A semigroup S is said to be stable if:

(a) $a, b \in S$ and $Sa \subset Sab$ implies that $Sa = Sab$; and

(b) $a, b \in S$ and $aS \subset baS$ implies that $aS = baS$.

3.31 Theorem. If S is either a compact or abelian semigroup, then S is stable.

Proof. If S is compact, then the fact that S is stable follows from the Swelling Lemma.

Suppose that S is an abelian semigroup, and let $a, b \in S$ such that $Sa \subset Sab$. Then $Sa \subset Sab = Sba \subset Sa$, and hence $Sa = Sab$. Similarly, $aS \subset baS$ implies that $aS = baS$. ■

Before proving our main result on stability, we digress to an interesting result which will be applied to a later example on the bicyclic semigroup.

3.32 Theorem. Let S be a compact semigroup and T a discrete subsemigroup of S. Then T is stable.

Proof. Let a, b ∈ T and suppose that $Ta \subset Tab$. Then $\overline{T}$ is a compact semigroup, $ab \in \overline{T}$ and $\overline{T}a \subset \overline{T}ab$. By virtue of 3.31, we see that $\overline{T}a = \overline{T}ab$. Let $t \in T$. Then $tab = xa$ for some $x \in \overline{T}$. Let $\{x_\alpha\}$ be a net in T such that $\{x_\alpha\} \to x$. Then $\{x_\alpha a\} \to xa = tab$ in T. Since T is discrete, we obtain that $x_{\alpha_0} a = tab$ for some α_0 and hence $tab \in Ta$. This proves that $Ta = Tab$. A similar argument works to show that $aT \subset baT$ implies that $aT = baT$. ∎

Observe that if S is a stable semigroup, then S^1 is stable. However, the converse is not generally true [O'Carroll, 1969].

3.33 Lemma. Let S be a stable semigroup and a, b ∈ S such that $L(a) \subset L(b) \subset J(a)$. Then $L(a) = L(b)$.

Proof. Since S is stable, so is S^1. Reformulating our hypothesis, we have $S^1a \subset S^1b \subset S^1aS^1$, so that $b = paq$ for some p, p ∈ S^1. Thus $S^1a \subset S^1b = S^1paq \subset S^1aq$. Since S^1 is stable, we have $S^1a = S^1aq$, and hence $S^1a = S^1b$, or equivalently, $L(a) = L(b)$. ∎

The next result appears in [Koch and Wallace, 1957].

3.34 Theorem. If S is a stable semigroup, then $\mathcal{D} = \mathcal{J}$.

Proof. By 3.28, we have that $\mathcal{D} \subset \mathcal{J}$.

To prove that $\mathcal{J} \subset \mathcal{D}$, let $(a,b) \in \mathcal{J}$. Then $S^1aS^1 = S^1bS^1$, and hence $a = xby$ for some x, y ∈ S^1. Now by $by \in S^1bS^1 = S^1aS^1$, so that $S^1a = S^1xby \subset S^1by \subset S^1aS^1$. By virtue of 3.33, we have $S^1a = S^1by$, and similarly $byS^1 = bS^1$. We conclude that $(b,by) \in \mathcal{R}$ and $(by,a) \in \mathcal{L}$. It follows that $(a,b) \in \mathcal{D}$, and the proof is complete. ∎

3.35 Corollary. If S is a compact semigroup, then $\mathcal{D} = \mathcal{J}$.

From what we have seen thus far in this section a $\mathcal{D}$-class of a semigroup is one of its basic building blocks (especially if the semigroup is compact).

It is convenient to think of a $\mathcal{D}$-class of a semigroup in terms of an "eggbox diagram" with $\mathcal{R}$-classes as rows and $\mathcal{L}$-classes as columns:

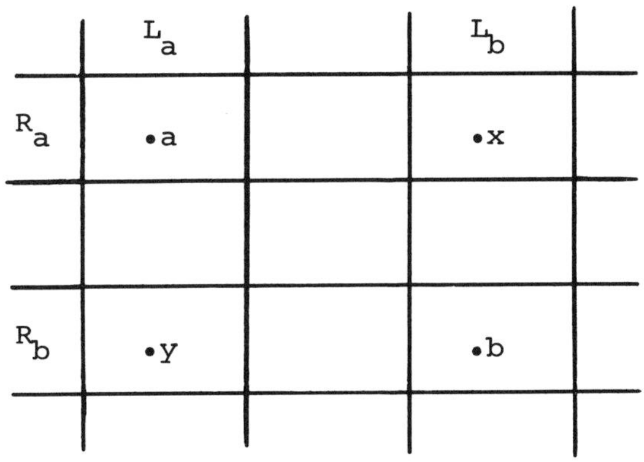

To say that a and b are in the same $\mathcal{D}$-class means that there exists $x \in R_a \cap L_b$, or equivalently there exists $y \in L_a \cap R_b$. Observe that the individual cells in the diagram above represent $\mathcal{H}$-classes.

A semigroup S is said to be $\mathcal{D}[\mathcal{L},\mathcal{R},\mathcal{J},\mathcal{H}]$ *simple* if S consists of a single $\mathcal{D}[\mathcal{L},\mathcal{R},\mathcal{J},\mathcal{H}]$ class.

3.36 Theorem. If S is a completely simple semigroup, then S is stable and $\mathcal{D}$-simple.

Proof. Let $S = [X,G,Y]_\sigma$. Suppose that $(a,g,b), (a',g',b') \in S$ with $S(a,g,b) \subset S(a,g,b)(a',g',b')$. From this inclusion we obtain that $b = b'$. Let $(x,u,y) \in S$ and $s = u\sigma(y,a)g\sigma(b,a')g'g^{-1}\sigma(y,a)^{-1}$. Then a straightforward computation shows that $(x,u,y)(a,g,b)(a',g',b) = (x,s,y)(a,g,b)$, so that $S(a,g,b) = S(a,g,b)(a',g',b')$. The verification of (b) in 3.10 is similar, and we conclude that S is stable. Since S is $\mathcal{J}$-simple, we have that S is $\mathcal{D}$-simple. ∎

3.37 Corollary. If S is a compact semigroup, then M(S) is $\mathcal{D}$-simple.

That a $\mathcal{D}$-simple semigroup need not be stable is illustrated with the classical bicyclic semigroup.

Let $X = \{p,q\}$ be a set consisting of two elements, F(X) the free semigroup generated by X, and adjoin an identity 1 to form $F(X)^1$. Let R be the congruence on $F(X)^1$ generated by $\{(pq,1)\}$ and let $S = F(X)^1/R$. The semigroup S is isomorphic to the bicyclic semigroup $\mathbb{B}$ (see the example following 1.17). If we identify p, q, and 1 with their images under the natural map of $F(X)^1$ onto S, then the elements of S can be displayed as follows:

$$\begin{array}{ccccc}
1 & q & q^2 & q^3 & \cdots \\
p & pq & pq^2 & pq^3 & \cdots \\
p^2 & p^2q & p^2q^2 & p^2q^3 & \cdots \\
p^3 & p^3q & p^3q^2 & p^3q^3 & \cdots \\
\vdots & \vdots & \vdots & \vdots &
\end{array}$$

The idempotents of S lie on the main diagonal and form an infinite descending chain in the ordering $e \leqslant f$ provided $e \in fSf$. The semigroup S is $\mathcal{D}$-simple with the rows forming the $\mathcal{R}$-classes and the columns forming the $\mathcal{L}$-classes. Observe that S is $\mathcal{H}$-trivial, i.e., each $\mathcal{H}$-class is degenerate. Since $pS \subset S = qpS$ and $pS \neq S$, we see that S is not stable. One can use this fact and 3.32 to show that no compact semigroup contains the bicyclic semigroup.

If a $\mathcal{D}$-class D of a semigroup S contains an idempotent, then, as we will see, each $a \in D$ has the property that $a \in aSa$, and in particular, there exist idempotents e and f such that $a \in ea$ and $a = af$. Thus for such $\mathcal{D}$-classes, we gain some insight into the multiplicative properties of the semigroup with regard to the elements of these $\mathcal{D}$-classes. Such $\mathcal{D}$-classes are called regular $\mathcal{D}$-classes.

An element a of a semigroup S is said to be *regular* if $a \in aSa$. If each element of a subset A of S is regular, then A is said to be *regular*.

3.38 Theorem. Let S be a semigroup and $a \in S$. Then these are equivalent:

(a) a is regular;

(b) D_a is regular;

(c) $R_a \cap E(S) \neq \square$; and

(d) $L_a \cap E(S) \neq \square$.

Proof. That (b) implies (a) is immediate.

To prove that (a) implies (b), suppose that a is regular and let $z \in D_a$. Then $a = axa$ for some $x \in S$, and $aS^1 = yS^1$ and $S^1y = S^1z$ for some $y \in S$. Consequently, we have $a = yu$, $y = av$, $z = ty$, and $y = sz$ for some $u, v, s, t \in S^1$. Finally, we obtain $z = ty = tav = taxav = t(yu)x(yu)v = (ty)uxyuv = zuxyuv = zux(sz)uv = zuxs(zuv) = zuxs(tyuv) = zuxs(tav) = zuxs(ty) = auxsz$, and hence Z is regular. This proves that D_a is regular.

To see that (a) implies (c), suppose that a is regular. Then $a = axa$ for some $x \in S$, so that $ax = axax$, and hence $ax \in E(S)$. Clearly, $axS^1 \subset aS^1$, and since $a = axa$, we have $a \in axS^1$. We obtain that $aS^1 \subset axS^1$, and $ax \in R_a \cap E(S)$.

To see that (c) implies (a), let $e \in R_a \cap E(S)$. Then $aS^1 = eS^1$, and hence $a = ex$ and $e = ay$ for some $x, y \in S^1$. From the second equation and the fact that $e = e^2$, we obtain $e = aye$, $a = ex = ayex = aya$, and a is regular.

The proof that (a) and (d) are equivalent is similar to the proof of the two preceding paragraphs. ∎

3.39 Corollary. Let D be a $\mathcal{D}$-class of a semigroup S. Then D is regular if and only if D contains an idempotent.

One can verify that the bicyclic semigroup is regular. However having observed that it is $\mathcal{D}$-simple, this is a consequence of 3.39.

3.40 Theorem. Let S be a subsemigroup of a semigroup T, and a and b elements of S which are regular in S, i.e., $a \in aSa$ and $b \in bSb$. Then:

(a) $(a,b) \in \mathcal{L}(S)$ if and only if $(a,b) \in \mathcal{L}(T)$;

(b) $(a,b) \in \mathcal{R}(S)$ if and only if $(a,b) \in \mathcal{R}(T)$; and

(c) $(a,b) \in \mathcal{H}(S)$ if and only if $(a,b) \in \mathcal{H}(T)$.

Proof. We will prove (b). The proof of (a) is similar, and (c) follows from (a) and (b).

To prove (b), suppose that $(a,b) \in \mathcal{R}(S)$. Then $aS^1 = bS^1 \subset bT^1$, and hence $a \in bT^1$. It follows that $aT^1 \subset bT^1$, and similarly $bT^1 \subset aT^1$. We obtain that $(a,b) \in \mathcal{R}(T)$.

On the other hand, suppose that $(a,b) \in \mathcal{R}(T)$. In view of 3.38, there exists $e \in R_a(S) \cap E(S)$ and $f \in R_b(S) \cap E(S)$. We have $eS^1 = aS^1$ and $fS^1 = bS^1$. By the preceding paragraph and the fact that $(a,b) \in \mathcal{R}(T)$, we see that $eT^1 = aT^1 = bT^1 = fT^1$. It follows that $e = fe \in fS^1$, so that $aS^1 = eS^1 \subset fS^1 = bS^1$, and similarly $bS^1 \subset aS^1$. We conclude that $(a,b) \in \mathcal{R}(S)$. ■

3.41 Theorem. Let S be a subsemigroup of a semigroup T and a, b ∈ S such that $(a,b) \in \mathcal{L}(S)$ $[\mathcal{R}(T), \mathcal{H}(T), \mathcal{D}(T), \mathcal{J}(T)]$. Then $(a,b) \in \mathcal{L}(T)$ $[\mathcal{R}(T), \mathcal{H}(T), \mathcal{D}(T), \mathcal{J}(T)]$.

Proof. This is straightforward. ■

The converse of 3.41 does not generally hold.

Let $T = \mathbb{D}$ the multiplicative unit complex disk, $0 < r < 1$, $a \neq b$ in T such that $|a| = |b| = r$. Let $S = \{z \in T : |z| \leqslant r\}$. Then $(a,b) \in \mathcal{D}(T)$. However, since $a \notin S^1 b$, we have that $(a,b) \notin \mathcal{D}(S)$.

3.42 Theorem. Let S be a semigroup, $e \in E(S)$, and $x \in eSe$. Then:

(a) $L_x(eSe) = L_x(S) \cap eSe$;

(b) $R_x(eSe) = R_x(S) \cap eSe$;

(c) $H_x(eSe) = H_x(S) \cap eSe$;

(d) $D_x(eSe) = D_x(S) \cap eSe$; and

(e) $J_x(eSe) = J_x(S) \cap eSe$.

Proof. We will prove (a). The proof of (b) is similar, and (c), (d) and (e) are consequences of (a) and (b).

To prove (a), let $y \in L_x(S) \cap eSe$. Then there exist $a, b \in S^1$ such that $zy = x$ and $bx = y$. Thus $x = exe$ and $y = eye$, and $x = ex = eay = eaeye = (eae)y$. Likewise, we obtain $y = (ebe)x$. It follows that $y \in L_x(eSe)$, and hence $L_x(S) \cap eSe \subset L_x(eSe)$. The other inclusion is obvious. ■

We turn now to the study of Green's relations on compact semigroups. Recall that we have already established that $\mathcal{D} = \mathcal{J}$ for compact semigroups.

3.43 Theorem. Let S be a compact semigroup. Then $\mathcal{L}$, $\mathcal{R}$, $\mathcal{D}$, and $\mathcal{H}$ are closed equivalence relations on S.

Proof. Again, we prove the result only for L.

Let $(a,b) \in S \times S \backslash L$. Then there exist disjoint open sets U and W such that $a \in U$ and $S^1 b \subset W$. By virtue of 1.1, there exists an open set V containing b such that $S^1 V \subset W$. It follows that $(a,b) \in U \times V \subset S \times S \backslash L$, and hence L is closed. The remainder of the conclusion follows from 3.25. ∎

3.44 Theorem. Let S be a compact semigroup with identity 1. Then $H(1) = H_1 = L_1 = R_1 = D_1$.

Proof. That $H(1) = H_1$ follows from 3.30. Let $x \in D_1 = J_1$. Then there exist $a, b \in S^1$ such that $axb = 1$. In view of 1.24, we see that $x \in H(1)$. It follows that $D_1 \subset H_1$. The conclusion now follows from 3.28. ∎

3.45 Corollary. Let S be a compact semigroup and $e \in E(S)$. Then $H(e) = H_e = L_e \cap eSe = R_e \cap eSe = D_e \cap eSe$.

Proof. This follows by applying 3.44 to the compact semigroup eSe, and then using 3.42. ∎

3.46 Theorem. Let S be a compact semigroup and $e \in E(S)$. Then

(a) $L_e = D_e \cap L(e)$; and

(b) $R_e = D_e \cap R(e)$.

Proof. We will prove (a). The proof of (b) is similar. To prove (a), observe that $L_e \subset D_e \cap L(e)$ is clear. Let $x \in D_e \cap L(e)$. Then $x \in S^1 e$ and $e = axc$ for $a, c \in S^1$. Since $x \in S^1 e$, we have that $x = xe$, and hence $e = a(xe)ce = ax(ece)$. Let $b = ece$. Then $e = axb$ and $b \in eSe$. By virtue of 1.24, we have that $b \in H_e = H(e)$. Let b^{-1} denote the inverse of b in H(e). Then $b^{-1} \in H_e = L_e \cap eSe$ (3.45). Observe that $b^{-1} = ax$. We obtain that $S^1 x \subset S^1 e = S^1 ax \subset S^1 x$. It follows that $S^1 x = S^1 e$, and hence $x \in L_e$. This completes the proof. ∎

We conclude this section with some simple examples.

Since the unit multiplicative complex disk $\mathbb{D}$ is abelian, we have $\mathcal{L} = \mathcal{R} = \mathcal{H} = \mathcal{D}$ is a congruence on $\mathbb{D}$. For $a \in \mathbb{D}$, we have $D_a = \{z \in \mathbb{D} : |z| = |z|\}$, and $\mathbb{D}/\mathcal{D}$ is topologically isomorphic to I_u.

Let S be the unit complex disk with multiplication $(a,b)\ a|b|$. Then for $z \in S$, we have $R_z = H_z = \{z\}$ and $L_z = D_z = \{t \in S : |t| = |z|\}$. If $(a,b) \in \mathcal{D}$ and $x \in S$, then $|a| = |b|$ and $|x|a|| = |x||a| = |x||b| = |x|b||$, so that $(xa,xb) \in \mathcal{D}$. It follows that $\mathcal{D}$ is a left congruence, and since $\mathcal{D} = \mathcal{L}$, we have that $\mathcal{D}$ is a right congruence. Thus $\mathcal{D}$ is a closed congruence on S, and it is not difficult to show that $S/\mathcal{D} \overset{T}{\approx} I_u$.

THE MULTIPLICATIVE STRUCTURE OF A $\mathcal{D}$-CLASS

As we have seen in the previous section, a $\mathcal{D}$-class of a semigroup S is one of its basic building blocks. This is especially true if S is a compact semigroup, since $\mathcal{D} = \mathcal{J}$ for compact semigroups. In this section we will investigate the multiplicative structure of a $\mathcal{D}$-class by employing Green's Translational Lemma (3.47) [Green, 1951]. There is an obvious dual to 3.47; which we will use but not state.

Observe that if S is a semigroup and $ax \in S$, then since $\mathcal{R}$ is a left congruence and $\mathcal{L}$ is a right congruence on S, we have that $aR_x \subset R_{ax}$ and $L_x a \subset L_{xa}$.

3.47 Theorem. Let S be a semigroup, $(x,y) \in \mathcal{L}$, and a, $b \in S^1$ such that $ax = y$ and $by = x$. Then

(a) $aR_x = R_y$;

(b) $bR_y = R_x$;

(c) $bat = t$ for each $t \in R_x$;

(d) $abt = t$ for each $t \in R_y$;

(e) $(au,u) \in \mathcal{L}$ for each $u \in R_x$; and

(f) $(bu,u) \in \mathcal{L}$ for each $u \in R_y$.

Proof. We will prove (a), (c) and (e). The proofs of (b), (d) and (f) are respectively similar.

To prove (c), let $t \in R_x$. Then $t = xw$ for some $w \in S^1$. We have, $bat = baxw = byw = xw = t$, and (c) is proved.

To prove (a), first observe that $aR_x \subset R_{ax} = R$. Let $z \in R_y$. Then $abz = z$ by (d), and $bz \in R_x$, since $bR_y \subset R_{by} = R_x$. We obtain that $z \in aR_x$, $R_y \subset aR_x$, and hence $aR_x = R_y$. This proves (a).

To prove (e), let $u \in R_x$. Then $S^1au \subset S^1u$ is clear, and $u = bau$ by (c), so that $S^1 \subset S^1au$. We obtain that $S^1au = S^1u$, or equivalently $(au,u) \in L$. ■

An alternate form of 3.47 states that $y_a|R_x : R_x \to R_y$ and $y_b|R_y : R_y \to R_x$ are mutually inverse and L-class preserving.

3.48 Corollary. Let S be a semigroup and $c, d \in S$.

(a) If $(c,dc) \in H$, then $dH_c = H_c$; and

(b) if $(c,cd) \in H$, then $H_cd = H_c$.

Proof. To prove (a), let $(c,dc) \in H$. We use 3.47 with $x = c$, $a = d$, $y = dc$, and $b \in S^1$ such that $bdc = c$. We first show that $dH_c \subset H_c$.

Let $t \in H_c$. Then $t \in R_c$, and hence $(dt,t) \in L$ (3.47 (e)). Thus $dt = L_t = L_c$. Now $dR_c = R_{dc} = R_c$ by 3.47 (a), and since $t \in R_c$, we have that $dt \in R_c$. It follows that $dt \in L_c \cap R_c = H_c$, and we proved that $dH_c \subset H_c$.

To prove that $H_c \subset dH_c$, let $z \in H_c = H_{dc}$. Then, since $z \in R_{dc}$, we have that $dbz = z$. We claim that $bz \in H_c$. Now $bz \in R_c$ by 3.47 (b). In view of 3.47 (f), we have $(bz,z) \in L$, and hence $bz \in L_z = L_c$. It follows that $bz \in R_c \cap L_c = H_c$. From $z = d(bz)$, we obtain that $z \in dH_c$, and hence $H_c \subset dH_c$. This completes the proof of (a). Part (b) is similar to (a) using the dual to 3.47. ■

3.47 Corollary. Let S be a semigroup and $e \in E(S)$.

(a) If $x \in L_e$, then $xR_e = R_x$ and $xH_e = H_x$; and

(b) if $x \in R_e$, then $L_ex = L_x$ and $H_ex = H_x$.

Proof. We prove (a) using 3.47. Again the proof of (b) is similar and uses the dual to 3.47.

To prove (a), first observe that $xR_e = R_x$ follows from 3.47 (b) by using $x = xe$ if $x \in L_e$.

Let $x \in L_e$. We want to show that $xH_e = H_x$. We will use 3.47 with $y = e$, $a \in S^1$ such that $ax = e$, and $b = x$, since $xe = x$.

To prove that $xH_e \subset H_x$, let $u \in H_e$. Then $u \in R_e$ and hence $xu \in R_x$ by 3.47 (b). Now since $u \in R_e$, we have $(xu,u) \in L$ by 3.47 (f). From $u \in H_e$, we have $u \in L_e = L_x$, so that $xu \in L_u = L_x$. Thus $xu \in L_x \cap R_x = H_x$, and we have proved that $xH_e \subset H_x$.

To prove that $H_x \subset xH_e$, let $v \in H_x$. Then $v \in R_x$ and $xav = v$ by 3.47 (c). We claim that $av \in H_e$. Now $v \in R_x$ implies that $(av,v) \in L$ by 3.47 (e). Thus $av \in L_v = L_x = L_e$. Again, since $v \in R_x$, we have that $av \in R_e$ by 3.47 (a). Finally, we have $av \in L_e \cap R_e = H_e$, and hence $v \in xH_e$. We conclude that $H_x \subset xH_e$, and the proof is complete. ■

3.50 Corollary. Let S be a semigroup and $x, y \in S$. Then these are equivalent:

(a) $xy \in H_y$;

(b) $xH_y \cap H_y \neq \square$; and

(c) $xH_y = H_y$.

Proof. That (a) implies (b), and (c) implies (a) is clear. To see that (b) implies (c), let $a \in H_y$ such that $xa \in H_y$. Then $(a,xa) \in H$, and hence $xH_a = H_a$ by 3.48 (a). This yields that $xH_y = H_y$, since $a \in H_y$, and the proof is complete. ■

3.51 Corollary. Let S be a semigroup and $x, y \in S$. If x, y and xy belong to the same H-class H_a, then H_a is a group.

Proof. Let $t \in H_a$. Now by 3.50, $sH_a = H_a$, and from $xt \in H_a$, we obtain $H_a = xH_a = H_a = H_{xt} = H_x t = H_a t$ using 3.48. We also have, $H_a y = H_a$, $ty \in H_a$, and $H_a = H_t y = H_{ty} = tH_y = tH_a$. It follows that $H_a t = H_a = tH_a$, and H_a is a group. ■

Observe from 3.51, that if H is an H-class in a semigroup, then either some pair of elements of H multiply to give an answer in H (in which case H is a group) or $H^2 \cap H = \square$. In particular, we have already seen, if H contains an idempotent, then H is a group.

3.52 Corollary. Let S be a semigroup and c, d $\in$ S. Then cd $\in R_c \cap L_d$ if and only if $R_d \cap L_c$ contains an idempotent (and hence $R_d \cap L_c$ is a group).

Proof. Suppose that cd $R_c \cap L_d$. Then $(c,d) \in \mathcal{L}$. We will apply 3.47 with x = d, y = cd, a = c, and b $\in S^1$ such that by = bcd = d. We claim that bc $\in R_d \cap L_c$. Now since c $\in R_{cd} = R_c$, we have $(bc,c) \in \mathcal{L}$ by 3.47 (f), and hence bc $\in L_c$. We also have that bc $\in R_d$ by 3.47 (a), since c $\in R_{cd}$. Thus bc $\in R_d \cap L_c$. In view of 3.47 (c), we have that bc(bc) = bc, and hence bc is the desired idempotent.

Suppose that $R_d \cap L_c$ contains an idempotent e. We want to show that cd $\in R_c \cap L_d$. Now $(c,e) \in \mathcal{L}$ and hence ce = c. We use 3.47 with x = c, y = e, b = c, and a $\in S^1$ such that ac $\in$ e. In view of 3.47 (f), since d $\in R_e$, we have $(cd,d) \in \mathcal{L}$, and hence cd $\in L_d$. We also have d $\in R_e$, and thus obtain cd $\in R_c$ by 3.47 (b). We conclude that cd $\in R_c \cap L_d$, and the proof is complete. ■

Suppose that $(c,d) \in \mathcal{D}$. Then in terms of an eggbox diagram, 3.52 can be interpreted as follows:

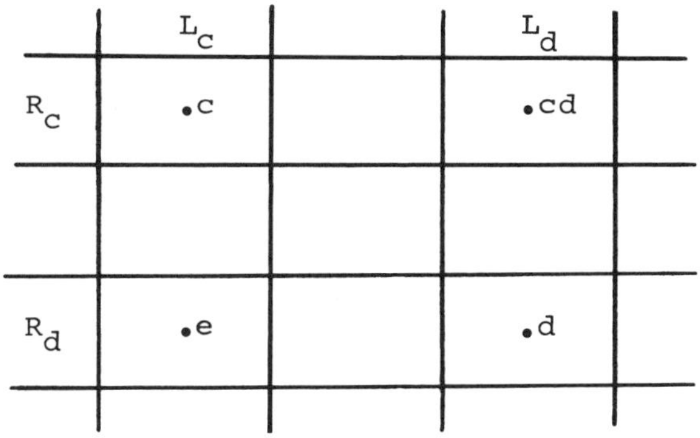

In other terms, if cd lands in one corner of the eggbox determined by c and d, then the other corner contains an idempotent, and vice versa.

3.53 Theorem. Let S be a semigroup with identity 1. Then each $\mathcal{D}$-class of S is invariant under inner automorphisms of

$H(1)$, i.e., $h^{1}xh \in D_x$ for each $h \in H(1)$ and $x \in S$, where h^{-1} denotes the inverse of h in $H(1)$.

Proof. Let $x \in S$ and $h \in H(1)$. In view of 3.30, we have that $H(1) = H_1 = L_1 \cap R_1$, by 3.28, so that $(1,h^{-1}) \in \mathcal{L}$. Since is a right congruence, we have $(x,h^{-1}x) \in \mathcal{L}$, and likewise $(xh,x) \in \mathcal{R}$. Using that $\mathcal{R}$ is a left congruence, we obtain $(h^{-1}xh,h^{-1}x) \in \mathcal{R}$, and hence $(h^{-1}xh,x) \in \mathcal{R} \circ \mathcal{L} = \mathcal{D}$. ∎

The following result appears in [Clifford and Miller, 1956].

3.54 Theorem. Let S be a semigroup and $x, y \in S$. Then $L_xR_y \subset D_{xy}$.

Proof. Let $a \in L_x$ and $b \in R_y$. Then $(a,x) \in \mathcal{L}$ and $(b,y) \in \mathcal{R}$. Since $\mathcal{L}$ is a right congruence and $\mathcal{R}$ is a left congruence, we obtain $(ab,xb) \in \mathcal{L}$ and $(xb,xy) \in \mathcal{L}$, and it follows that $(ab,xy) \in \mathcal{L} \circ \mathcal{R} = \mathcal{D}$. We conclude that $ab \in D_{xy}$, and $L_xR_y \subset D_{xy}$. ∎

SCHÜTZENBERGER GROUPS

The concept of a Schützenberger group was developed in [Schützenberger, 1957] and is useful in studying the relations between $\mathcal{H}$-classes within a $\mathcal{D}$-class, and particularly within a regular $\mathcal{D}$-class. We investigate the left Schützenberger group. There is, of course, a dual to this group called the right Schützenberger group which is anti-isomorphic to the left one.

3.55 Theorem. Let S be a semigroup, $a \in S$ $P(a) = \{x \in S^1 : xa \in H_a\}$, and $\mathcal{C}_a = \{(x,y) \in P(a) \times P(a) : xa = ya\}$. Then $P(a)$ is a subsemigroup of S^1, $\mathcal{C}_a$ is a congruence on $P(a)$, and $P(a)/\mathcal{C}_a$ is a group.

Proof. To see that $P(a)$ is a subsemigroup of S^1, let $u, v \in P(a)$. Then $ua \in H_a$ and $va \in H_a$. Thus $uva \in uH_a$. Since $ua \in H_a$, we have that $uH_a \subset H_a$, it follows that $uva \in H_a$, and $P(a)$ is a subsemigroup of S^1.

It is clear that $\mathcal{C}_a$ is an equivalence relation on $P(a)$, and that $\mathcal{C}_a$ is a left congruence. Suppose that $(x,y) \in \mathcal{C}_a$ and $z \in P(a)$.

Then $za \in H_a$, and in particular $(za,a) \in \mathcal{R}$. We obtain that $zaS^1 = aS^1$, so that $za = aw$ for some $w \in S^1$. It now follows that $xza = xaw = yaw = yza$, $(xz,yz) \in \mathcal{C}_a$, and $\mathcal{C}_a$ is a congruence on $P(a)$.

Let $G = P(a)/\mathcal{C}_a$ and $\pi : P(a) \to G$ be the natural map. It is straightforward to show that $\pi(1)$ is an identity for G. In view of the existence of an identity for G, in order to prove that G is a group, it is sufficient to prove that $G = tG$ for each $t \in G$. That $tG \subset G$ for each $t \in G$ is clear. Let $t \in G$ and $x \in P(a)$ such that $\pi(x) = t$, and let $u \in G$ with $u = \pi(y)$ for $y \in P(a)$. We will show that $u \in tG$. Now $(a,xa) \in \mathcal{H}$, since $x \in P(a)$. By virtue of 3.47, we have that $xH_a = H_a$, and hence $yz = xv$ for some $v \in H_a$, since $y \in P(a)$. Thus $(v,a) \in \mathcal{H} \subset \mathcal{L}$, and so $v = za$ for some $z \in S^1$. We obtain that $yz = xza$, and $z \in P(a)$, since $za = v \in H_a$. Finally, we have $u = \pi(y) = \pi(xz) = \pi(x)\pi(z) = t\pi(z) \in tG$, and $G \subset tG$. We conclude that G is a group. ∎

For the remainder of this section we adopt the notation introduced in 3.55.

If $a \in S$, then by 3.50, we have $P(a) = \{x \in S^1 : xa \in H_a\} = \{x \in S^1 : xH_a \cap H_a \neq \square\} = \{x \in S^1 : xH_a = H_a\}$. From $P(a) = \{x \in S^1 : xH_a = H_a\}$, we see that if $b \in H_a$ then $P(b) = \{x \in S^1 : xH_b = H_b\} = P(a)$, since $H_a = H_b$. Thus $P(a)$ does not depend on the particular element of H_a selected, but rather on the $\mathcal{H}$-class H_a itself. Likewise, if $x, y \in S^1$ and $xa = ya$, then we have that $b \in H_a$ implies that $b = au$ for some $u \in S^1$, so that $xb = xau = yau = yb$. It follows that if $(a,b) \in \mathcal{H}$, then $\mathcal{C}_a = \mathcal{C}_b$, and hence the group $P(a)/\mathcal{C}_a$ depends only on the $\mathcal{H}$-class H_a and not on the particular element selected.

If $a \in S$, then the group $P(a)/\mathcal{C}_a$ in 3.55 is called the Schützenberger group of $\mathcal{H}$-class H_a.

3.56 Theorem. Let S be a semigroup. Then the Schützenberger group is an invariant of a $\mathcal{D}$-class in S, i.e., if $(a,b) \in \mathcal{D}$ then $P(a)/\mathcal{C}_a$ is isomorphic to $P(b)/\mathcal{C}_b$.

Proof. Let $(a,b) \in \mathcal{D}$. Then there exists $c \in R_a \cap L_b$, so that $c = ax$, $a = cy$, $c = qb$, and $b = pc$ for some $x, y, p, q \in S^1$. Note that for $t \in P(a)$, we have that $ptqb = ptc = ptax \in pH_a x = pH_{ax} = pH_c = H_b$. Define $\phi : P(a) \to P(b)$ by $\phi(t) = ptq$ for $t \in P(a)$. Let $\alpha : P(a) \to P(a)/C_a$ and $\beta : P(a) \to P(a)/C_a$ be the natural maps.

Suppose that $t_1, t_2 \in P(a)$ such that $\alpha(t_1) = \alpha(t_2)$. Then $\phi(t_1)b = pt_1qb = pt_1c = pt_1ax = pt_2ax = pt_2c = pt_2qb = \phi(t_2)b$, and hence $(\phi(t_1), \phi(t_2)) \in C_b$. It follows that there exists a function $\phi^* : P(a)/C_a \to P(b)/C_b$ such that the diagram

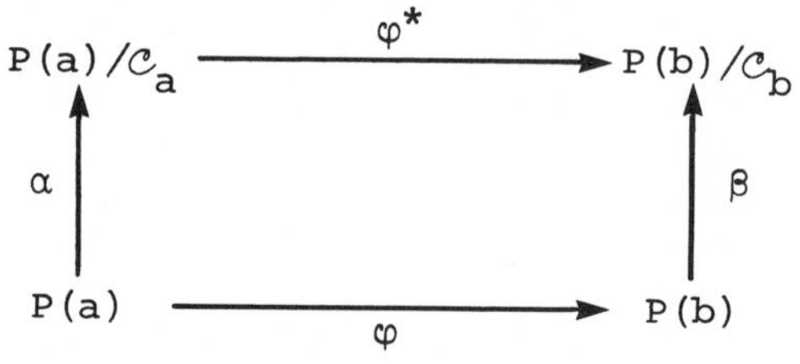

commutes. We will demonstrate that ϕ^* is an isomorphism.

To see that ϕ^* is injective, let $t_1, t_2 \in P(a)$, and suppose that $\phi^*\alpha(t_1) = \phi^*\alpha^*(t_2)$. Then $\beta(pt_1q) = \beta\phi(t_1) = \phi^*\alpha(t_1) = \phi^*\alpha(t_2) = \beta(pt_2q)$, so that $pt_1qb = pt_2qb$. We obtain that $t_1c = qpt_1qb = qpt_2qb = t_2c$, since qp acts as a left identity on cS. Therefore, $t_1a = t_1cy = t_2cy = t_2a$, and hence $\alpha(t_1) = \alpha(t_2)$. We obtain that ϕ^* is injective.

To see that ϕ^* is a homomorphism, let $t_1, t_2 \in P(a)$. We will show that $\phi^*(\alpha(t_1)\alpha(t_2) = \phi^*\alpha(t_1) \cdot \phi^*\alpha(t_2)$. Since α and β are homomorphisms, and the diagram above commutes, this reduces to showing that $\beta\phi(t_1t_2) = \beta(\phi(t_1) \cdot \phi(t_2))$. Note that $t_2qb = t_2c = t_2ax \in H_a x = H_{ax} = H_c$. Since qp acts as a left identity on cS, we put $pt_1(qp)t_2qb = pt_1t_2qb$, so that $\phi(t_1)\phi(t_2)b = \phi(t_1t_2)b$. It follows that $\beta(\phi(t_1)\phi(t_2)) = \beta\phi(t_1t_2)$, and we conclude that ϕ^* is a homomorphism.

To see that ϕ^* is surjective, we show that ϕ is surjective. Let $z \in P(b)$. Then $qzpa = pzpcy = pzby \in qH_b y = H_{qb}y = H_c y = H_{cy} = H_a$, and hence $qzp \in P(a)$. We obtain that $\phi(qzp) = pqzpq = z$ in view of 3.47, and its dual. It follows that ϕ is surjective, so that ϕ^* is surjective. This completes the proof of 3.56. ■

3.57 Theorem. Let S be a semigroup, and $a \in S$. Then H_a and $P(a)/\mathcal{C}_a$ have the same cardinal number.

Proof. Observe that if $t \in H_a$, then $t \in xa$ for some $x \in S^1$, and hence the function $\alpha : P(a) \to H_a$ defined by $\alpha(x) = xa$ is surjective. Let $\beta : P(a) \to P(a)/\mathcal{C}_a$ be the natural map. Then $\alpha(x) = \alpha(y)$ if and only if $\beta(x) = \beta(y)$ for each $x, y \in S^1$, and the conclusion follows. ∎

3.58 Corollary. Any pair of $\mathcal{H}$-classes within the same $\mathcal{D}$-class of a semigroup have the same cardinal number.

3.59 Theorem. Let S be a semigroup and $e \in E(S)$. Then H_e is isomorphic to $P(e)/\mathcal{C}_e$.

Proof. Define $\alpha : P(e) \to H_e$ by $\alpha(x) = xe$. Then α is surjective, as we have seen in the proof of 3.58. Let $x, y \in P(e)$, then $\alpha(xy) = xye = x(eye) = (xe)(ye) = \alpha(x)\alpha(y)$, and hence α is a homomorphism. If $\beta : P(e) \to P(e)/\mathcal{C}_e$ is the natural map, we see that $K_\alpha = K_\beta$, and hence $H_e \approx P(e)/\mathcal{C}_e$ by 1.49. ∎

3.60 Corollary. If S is a semigroup and $e, f \in E(S)$ such that $(e,f) \in \mathcal{D}$, then $H_e \approx H_f$.

The compact semigroup analog of the Schützenberger group was discussed in [Hofmann and Mostert, 1966]. We can obtain this analog by observing that if S is a compact semigroup and $a \in S$, then H_a is a compact subset of S, since $\mathcal{H}$ is closed in $S \times S$, and hence $P(a)$ is a compact subsemigroup of S^1. The congruence $\mathcal{C}_a$ on $P(a)$ is closed, and hence $P(a)/\mathcal{C}_a$ is a compact semigroup which is algebraically a group. It follows that $P(a)/\mathcal{C}_a$ is a compact group. If $(a,b) \in \mathcal{D}$, then the map ϕ in the diagram of the proof of 3.56 is clearly continuous, and since α is quotient, ϕ^* is continuous. We obtain that $P(a)/\mathcal{C}_a$ and $P(b)/\mathcal{C}_b$ are topologically isomorphic. Using the topological version of 1.49, we see that if S is compact and $e \in E(S)$, then $H_e \overset{T}{\approx} P(e)/\mathcal{C}_e$, and finally $e, f \in E(S)$ and $(e,f) \in \mathcal{D}$ implies that $H_e \overset{T}{\approx} H_f$. We summarize these results in the following:

3.61 Theorem. Let S be a compact semigroup.

(a) P(a) is a compact subsemigroup of S for each $a \in S$;

(b) $P(a)/\mathcal{C}$ is a compact group for each $a \in S$;

(c) if $(a,b) \in \mathcal{D}$, then $P(a)/\mathcal{C}_a \approx P(b)/\mathcal{C}_b$;

(d) if $e \in E(S)$, then $H_e \overset{\tau}{\approx} P(e)/\mathcal{C}_e$; and

(e) if $e, f \in E(S)$ and $(e,f) \in \mathcal{D}$, then $H_e \overset{\tau}{\approx} H_f$. ∎

GREEN'S QUASI-ORDERS

Here we investigate the various Green's quasi-orders, and how some conditions on a semigroup regarding these quasi-orders indicate some information about the structure of the semigroup. Material on Green's quasi-orders can be found in [Hofmann and Mostert, 1966], and [Wallace, 1953].

If S is a semigroup, then we define

(a) $\leqslant_L(S) = (a,b) \in S \times S : L(a) \subset L(b)$;

(b) $\leqslant_R(S) = (a,b) \in S \times S : R(a) \subset R(b)$;

(c) $\leqslant_J(S) = (a,b) \in S \times S : J(a) \subset J(b)$; and

(d) $\leqslant_H(S) = (a,b) \in S \times S : L(a) \subset L(b)$ and $R(a) \subset R(b)$.

If S is a semigroup and no confusion is likely, then we write $\leqslant_L$, $\leqslant_R$, $\leqslant_J$, and $\leqslant_H$ for $\leqslant_L(S)$, $\leqslant_R(S)$, $\leqslant_J(S)$, and $\leqslant_H(S)$, respectively. It is occasionally convenient to write $a \leqslant_L b$, $a \leqslant_R b$, $a \leqslant_J b$, and $a \leqslant_H b$ to mean $(a,b) \in \leqslant_L$, $(a,b) \in \leqslant_R$, $(a,b) \in \leqslant_J$, and $(a,b) \in \leqslant_H$, respectively.

It is useful to observe that if S is a semigroup and $a, b \in S$, then:

(a) $(a,b) \in \leqslant_L$ if and only if $a \in L(b)$;

(b) $(a,b) \in \leqslant_R$ if and only if $a \in R(b)$;

(c) $(a,b) \in \leqslant_J$ if and only if $a \in J(b)$; and

(d) $(a,b) \in \leqslant_H$ if and only if $a \in L(b) \cap R(b)$.

From the definition of Green's relations on a semigroup S, we see that $L = \leqslant_L \cap \leqslant_L^{-1}$, $R = \leqslant_R \cap \leqslant_R^{-1}$, $J = \leqslant_J \cap \leqslant_J^{-1}$, and $H = \leqslant_H \cap \leqslant_H^{-1}$.

3.62 Theorem. If S is a semigroup, then $\leqslant_H = \leqslant_L \cap \leqslant_R \subset \leqslant_L \cup \leqslant_R \subset \leqslant_J$.

Recall that a relation $\leqslant$ on a set X is a *quasi-order* on X provided $\leqslant$ is reflexive and transitive.

3.63 Theorem. If S is a semigroup, then $\leqslant_L$, $\leqslant_R$, $\leqslant_J$, and $\leqslant_H$ are quasi-orders on S.

The following theorem is useful in dealing with these relations:

3.64 Theorem. Let S be a semigroup and a, b, c $\in S^1$. Then:

(a) ab $\leqslant_L$ b;

(b) ab $\leqslant_R$ a;

(c) abc $\leqslant_J$ b; and

(d) aba $\leqslant_H$ a.

Proof. To prove (a), observe that $S^1ab \subset S^1b$, and hence ab $\leqslant_L$ b. The proof of (b) is similar to (a), and (c) and (d) follow from (a) and (b) using 3.62. ■

3.65 Theorem. Let S be a subsemigroup of a semigroup T and $\leqslant$ one of $\leqslant_L$, $\leqslant_R$, $\leqslant_J$, or $\leqslant_H$. If (a,b) $\in \leqslant$ (S), then (a,b) $\in \leqslant$ (T).

Proof. If (a,b) $\in \leqslant$ (S), then a $\in S^1b \subset T^1b$, and hence (a,b) $\in \leqslant_L$ (T). The proof for $\leqslant_R$, $\leqslant_J$, and $\leqslant_H$ is similar.

3.66 Theorem. Let S be a semigroup, e $\in$ E(S) and a, b $\in$ eSe:

(a) (a,b) $\in \leqslant_L$(eSe) if and only if (a,b) $\in \leqslant_L$(S);

(b) (a,b) $\in \leqslant_R$(eSe) if and only if (a,b) $\in \leqslant_R$(S);

(c) (a,b) $\in \leqslant_J$(eSe) if and only if (a,b) $\in \leqslant_J$(S); and

(d) (a,b) $\in \leqslant_H$(eSe) if and only if (a,b) $\in \leqslant_H$(S).

Proof. To prove (a), observe that (a,b) $\in \leqslant_L$(eSe) implies (a,b) $\in \leqslant_L$(S) follows from 3.65. Suppose, on the other hand, that (a,b) $\in \leqslant_L$(S). Then a = tb for some t $\in S^1$. We obtain that a = eae = etbe = et(ebe) = (ete)be = (ete)b, and hence a $\in (eS^1e)b$. It follows that (a,b) $\in \leqslant_L$(eSe), and (a) is proved. The proof of (b) follows that $(a,b) \in \leqslant_L(eSe)$, and (a) is proved. The proof of (b) is similar, and (c) and (d) follow from (a) and (b) and 3.62. ∎

Suppose that S is a semigroup and $e, f \in E(S)$. Then $e \leqslant_H f$ is equivalent to $e \in R(f) \cap L(f) = fS^1 \cap S^1f$, which in turn is equivalent to $e = fe = ef$. Thus the usual partial order on E(S) is the restriction of $\leqslant_H$ to E(S).

We turn to the study of $\leqslant_L$, $\leqslant_R$, $\leqslant_J$, and $\leqslant_H$ on compact semigroups.

3.67 Theorem. If S is a compact semigroup, then $\leqslant_L$, $\leqslant_R$, $\leqslant_J$, and $\leqslant$ are closed quasi-orders on S.

Proof. We present the proof for $\leqslant_L$. The proof for the other relations are similar.

To see that $\leqslant_L$ is closed, let $(a,b) \in S \times S \backslash \leqslant_L$. Then $a \in S \backslash S^1 b$ in view of 3.62, and hence there exist disjoint open sets U and W in S such that $a \in U$ and $S^1 b \subset W$. By virtue of 1.1, there exists an open set V containing b such that $S^1 V \subset W$. It follows that $(a,b) \in U \times V \subset S \times S \backslash \leqslant_L$, and $\leqslant_L$ is closed. The conclusion now follows from 3.63. ∎

3.68 Theorem. Let S be a compact semigroup and $\pi : S \times S/L[S/R,S/J,S/H]$ be the natural map of S onto the compact Hausdorff space $S/L[S/R,S/J,S/H]$. Then $\pi \times \pi(\leqslant_L)[\leqslant_R,\leqslant_J,\leqslant_H]$ is a closed partial order on $S/L[S/R,S/J,S/H]$.

Proof. We prove the result for $\pi : S \to S/L$. The proof is similar for the other relations.

That $\pi \times \pi(\leqslant_L)$ is a closed quasi-order follows from 3.67. We show that $\pi \times \pi(\leqslant_L)$ is symmetric. Suppose that (a,b) and (b,a) are in $\pi \times \pi(\leqslant_L)$. Let $a', b', a'', b'' \in S$ such that $a = \pi(a') = \pi(a'')$, $b = \pi(b') = \pi(b'')$, $(a',b') \in \leqslant_L$, and $(b'',a'') \in \leqslant_L$. Then $S^1a' \subset S^1b' = S^1b'' \subset S^1a'' = S^1a'$, $S^1a' = S^1b'$, and hence $a = \pi(a') = \pi(b') = b$. Thus $\pi \times \pi(\leqslant_L)$ is symmetric, and the proof is complete. ∎

If S is a semigroup such that for each $a \neq b$ in S, either $a \leqslant_L b$ or $b \leqslant_L a$ $[\leqslant_R,\leqslant_J,\leqslant_H]$, then S is said to be *totally* $L[R,J,H]$ *ordered*.

3.69 Theorem. Let S be a compact semigroup.

(a) If S is totally $\mathcal{L}$-ordered, then $\mathcal{L} = \mathcal{D}$ is a congruence on S;

(b) if S is totally $\mathcal{R}$-ordered, then $\mathcal{R} = \mathcal{D}$ is a congruence on S; and

(c) if S is totally $\mathcal{H}$-ordered, then $\mathcal{L} = \mathcal{R} = \mathcal{H} = \mathcal{D}$ is a congruence on S.

Proof. We will prove (a). The proof of (b) is dual to the proof of (a), and (c) following from (a) and (b) using 3.62.

Suppose that S is totally $\mathcal{L}$-ordered. We first show that $\mathcal{L} = \mathcal{D}$. Let $x \in S$. We claim that $R(x) \subset L(x)$.

Suppose that $R(x)\ L(x) \neq \square$. Let $y \in R(x)\ L(x)$. Then $y = xt$ for some $t \in S^1$, and since S is totally $\mathcal{L}$-ordered, we have $x \in L(y)$, so that $x = zy$ for some $z \in S^1$. We obtain that $S^1x = S^1zy \subset S^1y \subset S^1xt$. Now, by the Swelling Lemma, $S^1x = S^1xt$, and hence $L(x) = L(y)$, which contradicts $y \notin L(x)$. It follows that $R(x) \subset L(x)$ for each $x \in S$, and thus $J(x) \subset L(x) \subset J(x)$. We conclude that $J(x) = L(x)$, and $J_a = L_a$ for each $a \in S$. Since $\mathcal{J} = \mathcal{D}$ (315), we have $\mathcal{L} = \mathcal{D}$.

Since $\mathcal{L}$ is a right congruence, we complete the proof by showing that $\mathcal{L}$ is a left congruence. Suppose that $(a,b) \in \mathcal{L}$ and $x \in S$. Then $L(xa \subset S^1 \times S^1 = J(x) = L(x) = S^1x$, so that $L(xa) = L(x)$ by the Swelling Lemma, and similarly $L(xb) = L(x)$. We conclude that $L(xa) = L(xb)$, so that $(xa,xb) \in \mathcal{L}$, and $\mathcal{L}$ is a left congruence. ∎

If S is a semigroup, and $x \in S$, then $E_{\mathcal{L}}(x) = \{e \in E(S) : x \leqslant_{\mathcal{L}} e\}$, $E_{\mathcal{R}}(x) = \{e \in E(S) : x \leqslant_{\mathcal{R}} e\}$, $E_{\mathcal{J}}(x) = \{e \in E(S) : x \leqslant_{\mathcal{J}} e\}$, and $E_{\mathcal{H}}(x) = \{e \in E(S) : x \leqslant_{\mathcal{H}} e\}$.

If S is a semigroup and $x \in S$, then an element $e \in E_{\mathcal{L}}(x)$ $[E_{\mathcal{R}}(x), E_{\mathcal{J}}(x), E_{\mathcal{H}}(x)]$ is said to be a *minimal element* of $E_{\mathcal{L}}(x)$ $[E_{\mathcal{R}}(x), E_{\mathcal{J}}(x), E_{\mathcal{H}}(x)]$ if $f \in F_{\mathcal{L}}(x)$ $[E_{\mathcal{R}}(x), E_{\mathcal{J}}(x), E_{\mathcal{H}}(x)]$ and $f \leqslant_{\mathcal{L}} e$ $[\leqslant_{\mathcal{R}}, \leqslant_{\mathcal{J}}, \leqslant_{\mathcal{H}}]$ implies that $e = f$.

3.70 Theorem. Let S be a semigroup and $x \in S$.

(a) If $e \in E_{\mathcal{R}}(x)$, then $L_e x \subset L_x$;

(b) if $f \in E_{\mathcal{L}}(x)$, then $xR_f \subset R_x$; and

(c) if $a \in E_{\mathcal{R}}(x)$ and $f \in E_{\mathcal{L}}(x)$, then $L_e x R_f \subset D_x$ and $H_e x \cap xH_f \subset H_x$.

Proof. To prove (a), suppose that $e \in E_{\mathcal{R}}(x)$. Then $x \in R(e)$, and hence $ex = x$. Since $\mathcal{L}$ is a right congruence, $L_e x \subset L_{ex} = L_x$, and (a) is proved.

The proof of (b) is dual to the proof of (a).

To prove (c), suppose that $e \in E_{\mathcal{R}}(x)$ and $f \in E_{\mathcal{L}}(x)$. Then $ex = x$ and $xf = x$, so that $L_e x R_f \subset L_{ex} R_f = L_x R_f \subset D_{xf} = D_x$, using 3.55. Suppose $y \in H_e x \cap xH_f$. Then $y \in L_e x \subset L_{ex} = L_x$ and $y \in xR_f \subset R_{xf} = R_x$, so that $y \in L_x \cap R_x = H_x$, and $H_e x \cap xH_f \subset H_x$. ∎

3.71 Theorem. Let S be a compact semigroup and $x \in S$.

(a) If e is a minimal element of $E_{\mathcal{R}}(x)$, then $L_e x = L_x$;

(b) if f is a minimal element of $E_{\mathcal{L}}(x)$, then $xR_f = R_x$; and

(c) if e is a minimal element of $E_{\mathcal{R}}(x)$ and f is a minimal element of $E_{\mathcal{L}}(x)$, then $L_e x R_f = D_x$ and $H_e x \cap xH_f = H_x$.

Proof. To prove (a), suppose that e is a minimal element of $E_{\mathcal{R}}(x)$, and let $y \in L_x$. Then $x = ay$ and $y = bx$ for some $a, b \in S^1$. Since $e \in E_{\mathcal{R}}(x)$, we have that $ex = x$, and hence $x = (ea)y$ and $y = (be)x$. Let $z = eabe$. Then $zx = eabex = ea(bex) = eay = x$. Let f be the idempotent in $\Gamma(z)$. Then $x = z^n x$ for each $n \in \mathbb{N}$ (by induction) yields that $x = fx$. From this we see that $f \in Se$, since $z \in eSe$, so that $x \leq_{\mathcal{R}} f \leq_{\mathcal{R}} e$, and $f \in E_{\mathcal{R}}(x)$. By virtue of the minimality of e, we see that $e = f$. It follows that $z = eabe \quad H(e)$, $e \in S^1 be$, $be \in L_e$, and since $y = (be)x$, we conclude that $y \in L_e x$. Thus $L_x \subset L_e x$, and $L_e x \subset L_x$ (3.70), so that $L_x = L_e x$. This proves (a).

The proof of (b) is dual to the proof of (a).

To prove (c), suppose that e is minimal in $E_{\mathcal{R}}(x)$ and f is minimal in $E_{\mathcal{L}}(x)$. We will first prove that $L_e x R_f = D_x$. Now $L_e x R_f \subset D_x$ by 3.70, so we need to show that $D_x \subset L_e x R_f$.

Let $y \in D_x$. Then, since $\mathcal{D} = \mathcal{J}$ for S, there exist $a, b, c, d \in S^1$ such that $x = ayb$, and $y = cxd$. Now $ex = x = xf$, and hence

$x = eaybf$ and $y = cexfd$. Let $p = eace$ and $q = fdbf$. Then $x = pxq$ and $x = p^n xq^n$ for each $n \in \mathbb{N}$, inductively. Let g be the idempotent in $\Gamma(p)$ and h the idempotent in $\Gamma(q)$. Then $x = gxh$, $ege = g$, and $h = fhf$. In view of the minimality of e and f, we see that $e = g$, $h = f$, $p \in H(e)$, $q \in H(f)$, $c \in L_e$, $d \in R_f$, and since $y = cxd$, we have $y \in L_e x R_f$. It follows that $D_x \subset L_e x R_f$, and we conclude that $D_x = L_e x R_f$.

Now $H_e x \cap xH_f \subset H_x$ by 3.70. We will prove that $H_x \subset H_e x \cap xH_f$. Let $y \in H_x$. Then $y = xt$ for some $t \in S^1$, and since $ex = x$, we have that $ey = y$. We have that $y = ax$ and $x = by$ for some $a, b \in S^1$. It follows that $y = (eae)x$ and $x = (ebe)y$, so that $x = (ebe)y = (ebe)(eae)x$. Let $p = ebe$ and $q = eae$. Then $x = pqx$, and $pq \in eSe$. Let g be the idempotent in $\Gamma(pq)$. Then $g \in eSe$, $x = gx$, so that $g = e$ by the minimality of e. It follows that $pq \in H(e) \cap eSe$, so that $p, q \in H(e)$ by 1.17. Since $y = px$, we obtain that $y \in H_e x$. A similar argument works to show that $y \in xH_f$, and it follows that $y \in H_e x \cap xH_f$, and $H_x \subset H_e x \cap xH_f$. This completes the proof. ■

3.72 Theorem. Let S be a compact semigroup, $x \in S$, e a minimal element of $E_R(x)$, and f a minimal element of $E_L(x)$. Let $G_e = H_e \cap \{t \in S : tx \in H_x\}$ and $G_f = H_f \cap \{t \in S : xt \in H_x\}$. Then G_e is a compact subgroup of H_e, G_f is a compact subgroup of H_f, and $H_x = G_e x = xG_f$.

Proof. Now by 3.71, $H_x = H_e x \cap xH_f$, and $ex = x = xf$. Thus $ex \in H_x$, and G_e is a compact subsemigroup of H_e (see 3.55). It follows from 1.10 that G_e is a compact subgroup of H_e. Now $G_e x \subset H_x$ is clear from the definition of G_e. Now $H_x \subset H_e x$ implies that for $h \in H_x$, we get $h = gx$ for some $g \in H_e$, and hence by 3.48, $gH_x = H_x$. Thus $g \in G_e$, and $h = ge \in G_e x$. We conclude that $H_x \subset G_e x$, so that $H_x = G_e x$. The proof that G_f is a compact subgroup of H_f, and that $H_x = xG_f$ is similar. ■

Chapter 4

I-SEMIGROUPS

An I-semigroup is a semigroup on an arc such that one non-cut point is a zero and the other is an identity. Their structure was determined in [Mostert and Shields, 1957] using the earlier results in [Faucett, 1955], [Clifford, 1954], [Phillips, 1953], and [Cohen and Wade, 1953]. We will investigate the structure of I-semigroups and the closed congruences on I-semigroups. The congruences on I-semigroups were discussed in [Cohen and Krule, 1959].

TOPOLOGICAL PRELIMINARIES

Here we present some topological results which will be used in the sections which follow. No proofs are given in this section due to the abundant amount of background material required to do so. These results and their proofs appear in [Whyburn, 1963].

We use the word *continuum* to mean a compact connected Hausdorff space. A point p of a connected space X is called a *cut point* if $X\setminus\{p\}$ is not connected. It is well known that each non-degenerate continuum contains at least two non-cut points. An *arc* is a continuum with exactly two non-cut points. It is clear that the real unit interval [0,1] is an arc, and well known that any separable arc is homomorphic to [0,1].

Recall that a *total order* on a set X is a partial order $\leqslant$ on X such that for u and v in X either $u \leqslant v$ or $v < u$. We use $u \leqslant v$ to mean $u \leqslant v$ and $u \neq v$. For $x \in X$, let $x_U = \{y \in X : x \leqslant y\}$ and $x_L = \{y \in X : y \leqslant x\}$. Then $\mathcal{B} = \{x_L : x \in X\} \cup \{x_U : x \in X\}$ is a subbase for the closed sets of a topology on X called the *order topology*. For $x \leqslant y$ in X, we use [x,y] to denote $x_U \cap y_L$. The set X is said to be *order dense* if $u < w$ in X implies that $u < v < w$ for some $v \in X$, and X is said to be *order complete* if for each non-empty subset B of X, we have that sup B and inf B both exist.

Suppose that X is a non-degenerate continuum and that a and b are distinct non-cut points of X. Define $x < y$ in X if x is in the component of $X \setminus \{y\}$ containing a, and define $x \leqslant y$ if either $x < y$ or $x = y$. The relation $\leqslant$ is called the *cut point order* in X.

Recall that a connected space S is *irreducible connected* between a and b in X if no proper subset of X containing a and b is connected.

4.1 Theorem. Let S be a non-degenerate continuum, a and b distinct non-cut points of S, and $\leqslant$ the cut point order on X. Then these are equivalent:

(a) X is an arc;

(b) X is irreducibly connected between a and b; and

(c) $\leqslant$ is a total order on X such that X is order dense and order complete with $a = \inf X$ and $b = \sup X$, and X has the order topology.

If $(X,\leqslant)$ and $(Y,\leqslant)$ are partially ordered sets and $f : X \to Y$ is a function, then f is said to be *order-preserving* [*order-reversing*] if $x \leqslant y$ in X implies $f(x) \leqslant f(y)$ [$f(y) \leqslant f(x)$] in Y, and f is said to be *strictly order-preserving* [*order-reversing*] if f is injective and order-preserving [order-reversing].

4.2 Theorem. Let D be a dense subset of an arc [a,b] and let $g : D \to [0,1]$ be a strictly order-preserving dense function. Then there exists a homeomorphism $f : [a,b] \to [0,1]$ such that $f|D = g$.

4.3 Theorem. Let S be an arc, Y a non-degenerate Hausdorff space, and $f : X \to Y$ a continuous monotone function from X onto Y. Then Y is an arc.

BASIS I-SEMIGROUPS

Here we characterize semigroups on an arc $T = [a,b]$ with a acting as a zero and b as an identity, and with $E(T) = \{a,b\}$. There are exactly two such semigroups, I_u and I_n. This result and the other results in this section appear in [Faucett, 1955].

A topological semigroup on an arc [a,b] such that a is a zero and b is an identity is called an *I-semigroup*.

Throughout this section, let T = [a,b] denote the I-semigroup, and let $\leqslant$ denote the cut point order on T with a = inf T and b = sup T.

4.4 Lemma. If $x \in T$, then $xT = Tx = [a,x]$.

Proof. Since Tx is connected, $a = ax \in Tx$, and $x = bx \in Tx$, we have that $[a,x] \subset Tx$.

For the reverse inclusion, let $t \in T$ and consider tx. Since is a total order, we have $tx < x$ or $x \leqslant tx$. In the first case there is nothing to prove. In the latter case observe that $[a,x] \subset [a,tx] \subset t[a,x]$. By the Swelling Lemma, we obtain $[a,x] = [a,tx] = t[a,x]$. In view of the fact that [a,x] is an arc, we obtain that $x = tx$ and conclude that $[a,x] = Tx$. The argument that $[z,x] = xT$ is similar. ∎

4.5 Corollary. If $p, q \in T$, then $[a,p]q = [a,pq] = [a,p][a,q] = p[a,q]$.

Proof. By 4.4, $[a,p]q = [a,b]pq = [a,pq]$. ∎

4.6 Lemma. For each $x \in T$, λ_x and ρ_x are order preserving monotone functions.

Proof. Suppose that $u \leqslant v$ in T. Then $[a,ux] = [a,u]x \subset [a,v]x = [a,vx]$ and $ux \leqslant vx$, so that ρ_x is order preserving.

To see that ρ_x is monotone, fix $t \in T$ and consider $\rho_x^{-1}(t) = \{z \in T : xz = t\}$. Suppose that $z_1 < z_2$ in $\rho_x^{-1}(t)$, and choose z so that $z_1 \leqslant z \leqslant z_2$. Then $t = xz_1 \leqslant xz \leqslant xz_2 = t$, and hence $xz = t$. We obtain that $z \in \rho_x^{-1}(t)$, $[z_1,z_2] \subset \rho_x^{-1}(t)$, and $\rho_x^{-1}(t)$ is connected. ∎

4.7 Corollary. If $x, y, u, v \in T$ with $x \leqslant y$ and $u \leqslant v$, then $xu \leqslant yv$.

Proof. Since $x \leqslant y$, we have $xu \leqslant yu$, and since $u \leqslant v$, we have $yu \leqslant yv$. From the transitivity of $\leqslant$, we conclude that $xu \leqslant yv$. ∎

4.8 Corollary. If $x, y \in T$, then $xy \leqslant \min\{x,y\}$. In particular, $x^2 \leqslant x$.

Proof. We can assume that $x \leqslant y$ in T. Then $[a,xy] = x[a,y] \subset x[a,b] = [a,x]$, so that $xy \leqslant x = \min\{x,y\}$. ∎

If S is a semigroup and $x \in S$, then an element $y \in S$ is called a *square root* of x provided $y^2 = x$.

Observe that each element of a semigroup S has a square root if and only if the function $\phi : S \to S$ defined by $\phi(x) = x^2$ for each $x \in S$ is surjective.

4.9 Lemma. Each element of T has a square root.

Proof. Let $\phi : T \to T$ be the function $x \mapsto x^2$. Then ϕ is continuous by continuity of multiplication on T, and hence $\phi(T)$ is connected. In view of the fact that $\phi(a) = a$ and $\phi(b) = b$, we see that $\phi(T) = T$. ∎

If S is a semigroup with zero 0, then S is said to have *non-zero cancellation* provided $ux = uy \neq 0$ or $xu = yu \neq 0$ in S implies that $x = y$.

4.10 Lemma. If $E(T) = \{a,b\}$, then T has non-zero cancellation.

Proof. Suppose that $ux = uy \neq a$, and assume for the purpose of contradiction that $x < y$. Then by 4.4, we have $x = yt$ for some $t \in T$, and hence $ux = u(yt) = (uy)t$. Inductively, we obtain that $ux = uxt^n$ for each $n \in \mathbb{N}$, and hence $ux = uxe$, where e is the idempotent in the minimal ideal of $\Gamma(t)$. Now $t \neq b$, and $\{t^n\}$ is a non-increasing sequence by virtue of 4.4, so that $e = a$, since $E(T) = \{a,b\}$. We obtain that $ux = uxa = a$, which is a contradiction. ∎

4.11 Lemma. If $E(T) = \{a,b\}$, then each element of $T\backslash\{a\}$ has a unique square root.

Proof. Suppose that $u^2 = v^2 \neq a$ in T and that $u \leqslant v$. Then, by 4.7, we have $u^2 \leqslant uv \leqslant v^2$, and hence $u^2 = uv \neq a$. In view of 4.10, we see that $u = v$. The conclusion now follows from 4.9. ∎

We conclude this section with a characterization of those I-semigroups with exactly two idempotents.

4.12 Theorem. Suppose that $E(T) = \{a,b\}$.

(a) If T has no non-zero nilpotents, then $T \overset{\tau}{\approx} I_u$; and

(b) if T has non-zero nilpotents, then $T \overset{\tau}{\approx} I_n$.

Proof. To prove (a), let $x \in T$ such that $a < x < b$. By virtue of 4.11, there exists a unique sequence $\{x_n\}$ in T such that $x_1 = x$ and $x_n^2 = x_{n-1}$ for $n \geqslant 2$. Let Q_2 denote the set of all positive dyatic rational numbers. For $r \in Q_2$ with $r = p/2^b$ (p, $n \in \mathbb{N}$, p odd) define $x^r = x_n^p$. It is straightforward to prove that $x^r x^s = x^{r+s}$ for r, $x \in Q_2$, so that $D = \{x^r : r \in Q_2\}$ is a subsemigroup of T.

Observe that x^{2^n} for each $n \in \mathbb{N}$, and hence $x_n < b$ for each $n \in \mathbb{N}$. In view of 4.7, we see that $x_{n-1} = x_n^2 = x_n x_n \leqslant bx_n = x_n$ for $n \geqslant 2$, so that $\{x_n\}$ is an increasing sequence. Notice that we also have that for $r < s$ in Q_2, $x^s < x^r$ in D.

We claim that $\{x_n\} \to b$. Let $t = \sup \{x_n : n \in \mathbb{N}\}$. Then, since $\{x_n\}$ is increasing, we have $\{x_n\} \to t$. We obtain that $\{x_n^2\} \to t^2$, and since $x_n^2 = x_{n-1}$ for $n \geqslant 2$, we see that $t^2 = t$. By virtue of the fact that $E(T) = \{a,b\}$, we conclude that $t^2 = t$ and hence $t = b$.

We now establish that D is dense in T. Suppose, by way of contradiction, that $\overline{D} \neq T$, and let (u,v) be a component of $T \backslash \overline{D}$. (Note that (u,v) is open, since T is locally connected.) Now, since $\{x_n\} \to b$, we have that $\{x_n v\} \to b_v = v$, and $x_n v \leqslant bv = v$ for each $n \in \mathbb{N}$, by 4.7. If $x_n v = v$ for some $n \in \mathbb{N}$, then $x_n = b$ by 4.10, and we have a contradiction. Thus $x_n v < v$ for each $n \in \mathbb{N}$, and it follows that $u < x_m v < v$ for some $m \in \mathbb{N}$. Now $\overline{D}$ is a subsemigroup of T, and since v, $x_m \in \overline{D}$, we have $x_m v \in \overline{D}$, which contradicts $(u,v) \subset T \backslash \overline{D}$. We conclude that D is dense in T.

Define $g : D \to I_u$ by $g(x^r) = (1/2)^r$ for each $r \in Q_2$. Then g is a dense strictly order-reversing function. By 4.2, there exists a homeomorphism $f : T \to I_u$ such that $f|D = g$. Now g is a homomorphism of the dense subsemigroup D of T. Using this and the fact that

$f|D = g$, we obtain that f is a homomorphism. It follows that $f : T \to I_u$ is a topological isomorphism and the proof of (a) is complete.

The proof of (b) is similar to the proof of (a) with the following modifications: First, choose $x = \sup \{d \in T : d^2 = a\}$ and second, define $g : D \to I_n$ by $g(x^r) = t^r$ for $r \in Q_2$, where $t = \sup \{s \in I_n : s^2 = \frac{1}{2}\}$. ■

THE STRUCTURE OF I-SEMIGROUPS

The first result of this section appears in [Mostert and Shields, 1957].

4.13 Theorem. Let $S = [a,b]$ be an I-semigroup. If $e, f \in E(S)$, then $ef = \min \{e,f\}$. If $S \neq E(S)$, then $S \backslash E(S)$ is a union of disjoint open intervals. If I is the closure of one of these, then $I \overset{\tau}{\approx} I_u$ or $I \overset{\tau}{\approx} I_n$. Moreover, if $x \in I$ and $y \notin I$, then $xy = \min \{x,y\}$.

Proof. Let $e, f \in E(S)$ and suppose that $e \leqslant f$. In view of 4.7, we see that $e = e^2 \leqslant ef$. Now $f \leqslant b$, and again by 4.7, we have $ef \leqslant eb = e$. Since $\leqslant$ is a total order, we have $ef = e = \min \{e,f\}$.

Now suppose that $S \neq E(S)$ and let (e,f) be a component of $S \backslash E(S)$. Since $E(S)$ is closed, we have that $e, f \in E(S)$. Let $I = [e,f]$. Then, for $u, v \in I$, we have $e \leqslant u \leqslant f$, and $e \leqslant v \leqslant f$, and hence by 4.7, $e = e^2 \leqslant uv \leqslant f^2 = f$. We conclude that I is a subsemigroup of S.

To see that e is a zero for I, observe that for $u \leqslant f$, we have $[a,u] \subset [a,f] = f[a,b] \cap [a,b]f = fS \cup Sf$.

It is clear that $E(I) = \{e,f\}$, and hence by 4.12, either $I \overset{\tau}{\approx} I_u$ or $I \overset{\tau}{\approx} I_n$.

To establish the final assertion, suppose that $x \in I$ and $y \notin I$ with $x < y$. Then $e \leqslant x \leqslant f < y$, so that $e = ex \leqslant x^2 \leqslant xf = x \leqslant xy$. In view of 4.7, we have $xy \leqslant x$, and hence $xy = x = \min \{x,y\}$. ■

4.14 Corollary. Each I-semigroup is abelian.

CLOSED CONGRUENCES ON I-SEMIGROUPS

4.15 Lemma. Let S be an I-semigroup, R a closed congruence on S, $a \in S$, and $e = \sup \{f \in E(S) : f \leq a\}$.

(a) If $(e,a) \in R$, and $e \leq x \leq a$, then $(e,x) \in R$; and

(b) if there exists $z \in S$ such that $e \leq z < a$ and $(z,a) \in R$, then $(e,a) \in R$.

Proof. To prove (a), suppose that $(e,a) \in R$ and $e \leq x \leq a$. Then, by 4.4, $x = au$ for some $u \in S$ such that $e \leq u$. Since $(e,a) \in R$, we have $(eu,x) = (eu,au) \in R$. From 4.13, we have $eu = e$, so that $(e,x) \in R$.

To prove (b), suppose that $e \leq z \leq a$ and $(z,a) \in R$. Let $y = \inf \{t \in S : e \leq t \leq a \text{ and } (t,a) \in R\}$. Since R is closed, we have that $(y,a) \in R$. Suppose that $e < y$. Now $y = av$ for some $v \in S$ such that $y \leq v < a$ by 4.4 and 4.13. We obtain that $(yv,av) = (yv,y) \in R$ and hence $(yv,a) \in R$. However, $e \leq yv < y$, which contradicts the choice of y. It follows that $e = y$ and $(e,a) \in R$. ■

4.16 Theorem. Let S be an I-semigroup, R a closed congruence on S, $x \in S$, $a = \inf \{t \in S : (x,t) \in R\}$ and $b = \sup \{t \in S : (x,t) \in R\}$. Then $[a,b]$ is an R-class.

Proof. Since R is closed, we have that $(x,a) \in R$ and $(x,b) \in R$, and hence $(a,b) \in R$. It is clear that if $a = b$, then $[a,b] = \{x\}$ is an R-class. We assume then that $a < b$. Let $f = \inf \{t \in E(S) : a \leq t\}$. If $a \leq f \leq x$, then $(x,a) \in R$ yields that $(xf,af) = (f,a) \in R$, and hence by 4.15 (b), $(e,f) \in R$, where $e = \sup \{t \in E(S) : t \leq a\}$. It follows that $a \in E(S)$. If $a \leq x \leq f$, then $a \in E(S)$ by 4.15 (a). Let $y \in S$ such that $a < y < b$. Then $y = bu$ for some $u \in S$ such that $a \leq u$ by 4.4. Since $(a,b) \in R$, we have $(a,y) = (au,bu) \in R$ and it follows that $[a,b]$ is an R-class. ■

Observe in the proof of 4.16 that if P is a non-degenerate R-class, then $P = [c,b]$ for some $e \in E(S)$ and $b \in S$ such that $e \leq b$. In view of this fact we obtain:

4.17 Corollary. Let S be an I-semigroup, R a closed congruence on S, $x \in S$, $e = \sup \{t \in E(S) : t \leqslant x\}$, $f = \inf \{t \in E(S) : x \leqslant t\}$ and T the subsemigroup [e,f]. Then there exists a closed ideal J of T such that $R \cap (T \times T) = (J \times J) \cup \Delta(T)$.

The following result appears in [Cohen and Krule, 1959]:

4.18 Theorem. Let S be an I-semigroup and R a closed congruence on S such that S/R is a non-degenerate. Then S/R is an I-semigroup.

Proof. Let $\pi : S \to S/R$ denote the natural map. In view of 4.6, we see that π is monotone. The conclusion follows from 4.3. ∎

Combining 1.49 with 4.18 we have:

4.19 Corollary. A continuous homomorphic image of an I-semigroup is either degenerate or an I-semigroup.

As a consequence of 4.17 we have:

4.20 Corollary. If S is a non-degenerate continuous homomorphic image of I_u, then either $S \overset{\tau}{\approx} I_u$ or $S \overset{\tau}{\approx} I_n$.

4.21 Corollary. If S is a non-degenerate continuous homomorphic image of I_n, then $S \overset{\tau}{\approx} I_n$.

4.22 Corollary. If S is a non-degenerate continuous homomorphic image of I_m, then $S \approx I_m$.

We employ the results here to obtain the theorem:

4.23 Theorem. Let S be a compact semigroup, T an I-semigroup in S, and $a \in S$ such that aT [Ta] is non-degenerate. Then aT [Ta] is an arc.

Proof. Observe that $\lambda_a : T \to aT$ is a continuous function from T onto aT. Let $R = \{(x,y) \in T \times T : \lambda_a(x) = \lambda_a(y)\}$. Then R is a closed congruence on T. Let $\pi : T \to T/R$ denote the natural map. In view of the fact that π is a quotient map, we obtain a homeomorphism $\phi : T/R \to aT$ such that the diagram:

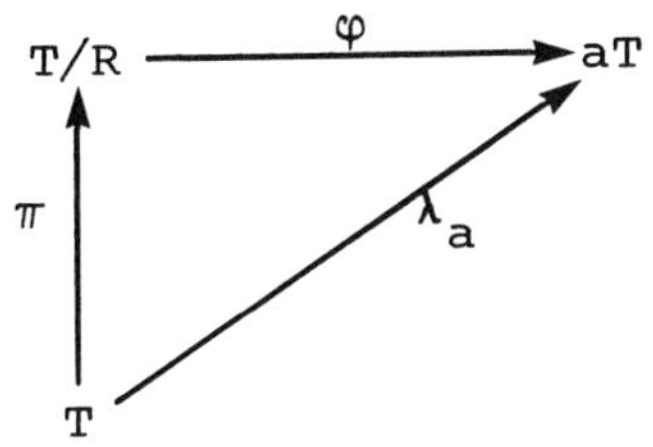

commutes. By 4.18, we have T/R is an I-semigroup, and hence aT is an arc. ■

We close this chapter with the following example:

Let $S = I_u \times I_u$, $T = \{(x,1) : 0 \leqslant x \leqslant 1\}$, and $a = (1/2,1/2)$. Then $T \overset{\tau}{\approx} I_u$, and $aT = Ta = \{(x,1/2) : 0 \leqslant x \leqslant 1/2\}$ is an arc in S.

Chapter 5

ONE PARAMETER SEMIGROUPS

The main result of this chapter is the existence of a one parameter semigroup leaving H(1) in a compact connected monoid S having no other idempotents near 1. We depart from the original proof in [Mostert and Shields, 1957] and follow the outline given in [Carruth and Lawson, 1970b]. We begin with some preliminary results.

DIVISIBILITY AND CONNECTEDNESS

Here we introduce the concept of divisibility in a semigroup. Our primary purpose for its introduction at this point is to establish a result that will be employed in the proof of the theorem on existence of one-parameter semigroups; namely that if S is a compact divisible semigroup and $e \in E$, then H(e) is connected. Compact divisible semigroups will be treated in more detail in Chapter 6.

A semigroup S is said to be *divisible* if for each $y \in S$ and $n \in \mathbb{N}$, there exists $x \in S$ such that $x^n = y$.

It is clear that the surmorphic image of a divisible semigroup is divisible, and that the cartesian product of divisible semigroups is divisible.

5.1 Lemma. If S is a finite divisible semigroup, then $S = E$.

Proof. Let S be a finite divisible semigroup, and $a \in S$. Then $\Gamma(a)$ is finite and contains an idempotent. Thus some power of a is idempotent. Let $x_1, \ldots, x_m$ denote the elements of S, and for each j $(1 \leq j \leq m)$ let $n_j \in \mathbb{N}$ such that $x_j^{n_j} \in E$. Let $n = n_1 n_2 \cdots n_m$. Then

$a^n \in E$ for each $a \in S$. Since S is divisible, the function $x \mapsto x^n$ from S into S is surjective. We conclude that S = E. ■

5.2 Lemma. If S is a compact totally disconnected divisible semigroup, then S = E.

Proof. In view of 2.26, we see that there exists a collection $\{F_\alpha : \alpha \in A\}$ of finite semigroups and bonding homomorphisms such that S is the strict projective limit of the F_α's, and hence each projection $\pi_\alpha : S \to F$ is surjective. It follows that each F_α is divisible, and hence by 5.1 consists of idempotent elements. Since S is a subsemigroup of $\Pi\{F_\alpha : \alpha \in A\}$, we conclude that S = E. ■

5.3 Theorem. Let S be a compact semigroup. Then S is divisible if and only if each component of S is a divisible subsemigroup of S.

Proof. If each component of S is a divisible subsemigroup of S, then it is immediate that S is divisible.

Suppose, on the other hand, that S is divisible. Consider the monotone-light factorization of the map ϕ sending S to a degenerate semigroup 1:

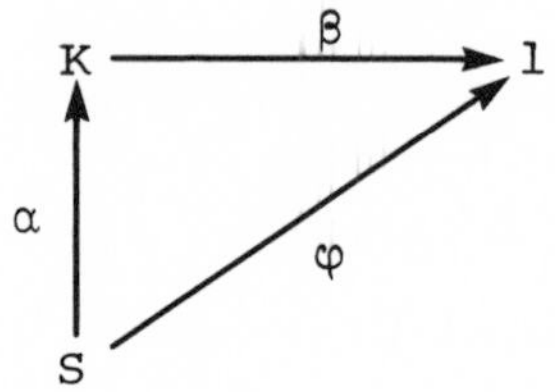

Then, since α is a surmorphism, K is divisible and hence a compact totally disconnected divisible semigroup. From 5.2, we have that K = E(K). It is clear that each component of S is of the form $\alpha^{-1}(e)$ for $e \in K$, and hence is a subsemigroup of S. Thus for $x \in S$ and $n \in \mathbb{N}$, we have that x and x^n are in the same component of S. The conclusion now follows. ■

5.4 Theorem. If S is a divisible semigroup and $e \in E$, then H(e) is a divisible subgroup of S.

Proof. Let $x \in H(e)$ and $n > 1$ in $\mathbb{N}$. Then, since S is divisible, $y^n = x$ for some $y \in S$. Now $xy = y^n y = n^{y+1} = yy^n = yx$, so that x commutes with y. Let x^{-1} denote the inverse of x in H(e). Then $eye = x^{-1}xye = x^{-1}yxe = x^{-1}yx = x^{-1}xy = ey$, and $eye = eyxx^{-1} = exyx^{-1} = xyx^{-1} = yxx^{-1} = ye$, so that $ey = ye$. We obtain that $(eye)^n = (eye)\cdots(eye) = ey^n = ex = x$. Now $(eye)(eye)^{n-1} = x$ and $eye[(eye)^{n-1}x^{-1}] = e$, so that $e \in eyeS$, and likewise $e \in Seye$. We obtain that $eye \in H(e)$, $(eye)^n = x$, and hence H(e) is divisible. ∎

<u>*5.5 Theorem*</u>. Let S be a compact divisible semigroup and $e \in E$. Then H(e) is connected. In particular, each compact divisible group is connected.

Proof. Now by 5.4, H(e) is divisible, and hence by 5.3, each component of H(e) is a compact divisible subsemigroup. Thus each component of H(e) contains an idempotent. Since e is the only idempotent in H(e), we conclude that H(e) is connected. ∎

It is established in [Mycielski, 1958] that divisibility and connectedness are equivalent for a compact group. However, 5.5 is sufficient for our purpose.

EXISTENCE OF ONE-PARAMETER SEMIGROUPS

A *one-parameter semigroup* in a topological semigroup S is a continuous homomorphism $\sigma : \mathbb{H} \to S$. A continuous injective function $\tau : [0,1] \to S$ is called a *local one-parameter semigroup* provided $\tau(a + b) = \tau(a)\tau(b)$, whenever $a, b, a + b \in [0,1]$.

<u>*5.6 Lemma*</u>. Let S be a topological semigroup and let $\tau : [0,1] \to S$ be a local one-parameter semigroup. Then there exists a unique one-parameter semigroup $\sigma : \mathbb{H} \to S$ such that $\sigma|[0,1] = \tau$.

Proof. For $x \in \mathbb{H}$, let $n \in \mathbb{N}$ and $p \in [0,1]$ such that $x = np$. Define $\sigma(x) = \tau(p)^n$. Then it is tedious but straightforward to prove that σ is the desired one-parameter semigroup, and that the condition that $\sigma|[0,1] = \tau$ holds and ensures the uniqueness of σ. ∎

Before establishing the main theorem of this chapter, we consider some preliminary definitions and observations.

If $(X,\mathcal{U})$ is a uniform space, $A \subset X$, and $U \in \mathcal{U}$, then A is said to be U-*connected* provided that for each a, b $\in$ A, there exists $x_0,\ldots,x_n \in A$ with $x_0 = a$, $x_n = b$, and $(x_i,x_{i+1}) \in U$ for $i = 0,1,\ldots,n-1$. A discussion of this concept appears in [Hofmann and Mostert, 1966].

A net $\{x_\alpha\}$ in a set X is said to be a *universal net* in X if for each $A \subset X$, $\{x_\alpha\}$ is eventually in A or $X\backslash A$. Universal nets are discussed in [Kelly, 1955].

If S is a compact monoid (1 = identity of S), then a subsemigroup T of S is said to be *irreducible* provided:

(a) T is a compact connected subsemigroup of S;

(b) $1 \in T$;

(c) $T \cap M(S) \neq \square$; and

(d) T is minimal with respect to (a), (b), and (c).

In a compact Hausdorff space X, the neighborhoods of Δ (the diagonal of X) form a uniform structure on X whose uniform topology is the given topology on X.

5.7 Lemma. Let S be a compact semigroup and let $\mathcal{U}$ be an open subset of $S \times S$ containing Δ. Define $\mathcal{V} = \{(x,y) : (x,y) \cup \Delta(x,y) \cup (x,y)\Delta \cup \Delta(x,y)\Delta \subset \mathcal{U}\}$. Then $\mathcal{V}$ is an open Δ-ideal and $\Delta \subset \mathcal{V} \subset \mathcal{U}$.

Proof. Clearly $\Delta \subset \mathcal{V} \subset \mathcal{U}$ and $\mathcal{V}$ is a Δ-ideal. That $\mathcal{V}$ is open follows from 1.1. ■

If G is a topological group, then a continuous homomorphism $\chi : G \to \mathbb{R}/\mathbb{Z}$ is called a *character of* G. To say that χ is non-trivial means that $\chi(G) \neq \{1\}$, where 1 is the identity of $\mathbb{R}/\mathbb{Z}$.

We will employ the following significant result. It is proved in [Pontryagin, 1966].

5.8 Lemma. A non-degenerate compact abelian group has a non-trivial character.

In fact, 5.8 holds if "compact" is replaced by "locally compact". We remark that 5.8 is the only result we call upon, for the proof of the main theorem, whose proof is not immediately accessible in this book.

To continue the construction of the machinery we will use in the proof of the main theorem, we recall three results from general topology.

5.9 Lemma. Let X and Y be Hausdorff spaces.

(a) If $f : X \to Y$ is a function, $A \subset X$, and $B \subset Y$, then $f(A \cap f^{-1}(B)) = f(A) \cap B$;

(b) If X is compact, Y is connected, $f : X \to Y$ is a continuous open surjection, and M is a component of X, then $f(M) = Y$. In particular, if A is a connected subset of X and for some $y \in Y$ we have $f^{-1}(y) \subset A$, then X is connected; and

(c) If R is a closed equivalence relation on X, $\rho : X \to X/R$ is the natural quotient map, and if B is either an open or closed subset of X/R, then B is homeomorphic to $\rho^{-1}(B)/R_0$, where $R_0 = R \cap (\rho^{-1}(B) \times \rho^{-1}(B))$.

The next result is a variation of a result appearing in [Yamabe, 1953].

5.10 Lemma. Let T be a compact monoid, U an open subset of T containing 1 such that $\overline{U}$ contains no non-trivial subsemigroup, and let W be a neighborhood of 1 such that $W \subset \overline{W} \subset U$. Then there exists an open set V containing 1 such that $V \subset W$ and such that if $x \in V$ and $n \in \mathbb{N}$, then the condition that $x^k \in U$ for all $k \in \mathbb{N}$ with $1 \leqslant k \leqslant n$ and $x^n \in V$ implies that $x^k \in W$ for all $k \in \mathbb{N}$ with $1 \leqslant k \leqslant n$.

Proof. Let $\{V_\alpha : \alpha \in D\}$ be a neighborhood base at 1 such that $V_\alpha \subset W$ for all $\alpha \in D$ and $\alpha \leqslant \beta$ in D implies that $V_\beta \subset V_\alpha$.

We establish the result using proof by contradiction. For this purpose, assume that for each $\alpha \in D$, there exists $x_\alpha \in V_\alpha$, $n_\alpha \in \mathbb{N}$, and $k_\alpha \in \mathbb{N}$ with $k_\alpha < n_\alpha$ such that $x_\alpha^m \in U$ when $m \in \mathbb{N}$ and $1 \leqslant m \leqslant n_\alpha$, $x_\alpha^{n_\alpha} \in V_\alpha$, but $x_\alpha^{k_\alpha} \notin W$. Then $\{x_\alpha^{k_\alpha}\}$ clusters to some x in $\overline{U}\backslash W$. In particular, $x \neq 1$. By considering subnets, we can assume that $\{x_\alpha^{k_\alpha}\} \to x$. We will arrive at our contradiction by showing that $\{1,x,x^2,\ldots\} \subset \overline{U}$. Let $m \in \mathbb{N}$. Then there exists m_α and r_α in $\mathbb{N} \cup \{0\}$ such that $mk_\alpha = m_\alpha n_\alpha + r_\alpha$, $r_\alpha < n_\alpha$, and $m_\alpha \leqslant m$ for each $\alpha \in D$. Now let P be an open set containing 1. Then for each $\beta \in D$, there exists $\alpha \geqslant \beta$ such that $V_\alpha^m \subset P$. For such an α, we have $x_\alpha^{n_\alpha} \in V_\alpha$ and $x_\alpha^{m_\alpha n_\alpha} = (x_\alpha^{n_\alpha})^{m_\alpha} \in V_\alpha^{m_\alpha} \subset V_\alpha^m \subset P$. Hence the net $\{x_\alpha^{m_\alpha k_\alpha}\}$ clusters to 1. Again, we can assume that $\{x_\alpha^{m_\alpha k_\alpha}\} \to 1$. Now if $r_\alpha < n_\alpha$ in $\mathbb{N}$, then $x_\alpha^{r_\alpha} \in V$, so that $\{x_\alpha^{r_\alpha}\}$ clusters to some y in $\overline{U}$. We assume that $\{x_\alpha^{r_\alpha}\} \to y$. Finally, $mk_\alpha = m_\alpha n_\alpha + r_\alpha$ implies that $x^{mk_\alpha} = x_\alpha^{r_\alpha} x_\alpha^{m_\alpha n_\alpha}$ for each α, $\{x_\alpha^{mk_\alpha}\} = \{(x_\alpha^{k_\alpha})^m\} \to x^m$, and $\{x_\alpha^{r_\alpha} x_\alpha^{m_\alpha n_\alpha}\} \to y \cdot 1 = y \in \overline{U}$. We conclude that $y = x^m$, $x^m \in \overline{U}$, and $\{1,x,x^2,\ldots\} \subset \overline{U}$. This contradicts the hypothesis that $\overline{U}$ contains no non-trivial subsemigroup of T. ∎

5.11 Lemma. Let S be a compact connected monoid. Then S contains an irreducible subsemigroup which contains 1_S and meets M(A).

Proof. Consider the collection $\mathcal{K}$ of all compact connected subsemigroups K of S such that $1 \in K$ and $K \cap M(S) \neq \square$. Then $\mathcal{K}$ is partially ordered by inclusion. Let $\mathcal{C}$ be a maximal chain in $\mathcal{K}$ (using the Hausdorff Maximality Principal). Then $T = \cap\, \mathcal{C}$ is an irreducible subsemigroup of S having the desired projection. ∎

5.12 Lemma. Let T be an irreducible abelian semigroup. Then T is divisible.

Proof. The map $x \to x^n$ is a continuous homomorphism for each $n \in \mathbb{N}$. ∎

If X is a space and $\mathcal{U}$ is a neighborhood of Δ (the diagonal of X), and $A \subset X$, we let $\mathcal{U}[A] = \{y \in X : (x,y) \in \mathcal{U}$ for some $x \in A\}$.

If the H quasi-order $\leqslant_H$ on a semigroup S is a total quasi-order; i.e., $\leqslant_H$ is reflexive and transitive and either $x \leqslant_H y$ or $y \leqslant_H x$ for $x, y \in S$; then S is said to be an *H-chain*.

If $\{A_\alpha\}$ is a net of subsets of a space X, then $\lim\sup A_\alpha = \{y \in X :$ for each open set U containing y, $U \cap A_\alpha$ is frequently non-empty$\}$ and $\lim\inf A_\alpha = \{y \in X :$ for each open set U containing y, $U \cap A_\alpha$ is eventually non-empty$\}$. We call $\lim\sup A_\alpha$ the *limit superior* of the net $\{A_\alpha\}$, and $\lim\inf A_\alpha$ is called the *limit inferior* of the net $\{A_\alpha\}$. It is immediate that $\lim\inf A_\alpha \subset \lim\sup A_\alpha$. If $\lim\inf A_\alpha = \lim\sup A_\alpha$, then we let $\lim A_\alpha = \lim\inf A_\alpha = \lim\sup A_\alpha$, and call $\lim A_\alpha$ the *limit* of the net $\{A_\alpha\}$. In a Hausdorff space X, we have that $\lim\sup A$ and $\lim\inf A$ are closed in X. Hence, if $\lim A_\alpha$ exists, it is closed in X.

We are now prepared to prove the main theorem of this section:

> *5.13 Theorem.* Let S be a compact monoid such that H(1) is not open in S and there exists an open neighborhood V of 1 such that $E \cap V = \{1\}$. Then there exists a one-parameter semigroup $\sigma : \mathbb{H} \to S$ such that $\sigma(0) = 1$ and $\sigma(a) \notin H(1)$ if $a > 0$.

Proof. The existence of a local one-parameter semigroup $\sigma : [0,1] \to S$ such that $\sigma(a) \in H(1)$ for $0 \leqslant a \leqslant 1$ will be established. The conclusion will then follow from 5.6.

We will establish the existence of the desired local one-parameter semigroup by using a sequence of reductions.

I. Reduce to the Case E = {0,1}.

We claim that $S(E\backslash\{1\})S$ is a closed ideal containing $E\backslash\{1\}$. To see this, observe that $\{1\}$ is open and closed in E. Hence $E\backslash\{1\}$ is closed and is therefore compact. We obtain that $S(E\backslash\{1\})S$ is a closed ideal. Clearly, $E\backslash\{1\} \subset S(E\backslash\{1\})S$.

Now let $T = S/S(E\backslash\{1\})S$ and let $\phi : S \to T$ be the natural quotient homomorphism.

We first claim that $E(T) = \{0,1\}$. Clearly, $\{0,1\} \subset E(T)$. Now if $e \in S$, and $\phi(e)^2 = \phi(e)$, then $(e^2,e) \in \Delta_S \cup [S(E\backslash\{1\})S \times S(E\backslash\{1\})S]$, since $\phi(e)^2 = \phi(e^2)$. If $(e^2,e) \in \Delta_S$, then either $e = 1$ or $e \in E\backslash\{1\}$, so that either $\phi(e) = 1$ or $\phi(e) = 0$. If $(e^2,e) \in S(E\backslash\{1\})S \times S(E\backslash\{1\})S$, then $\phi(e) = 0$. In either situation, we conclude that $E(T) = \{0,1\}$.

We next claim that $S(E\backslash\{1\})S \cap H(1) = \square$. To see this, first note that if a subgroup of a semigroup meets an ideal, then the ideal contains the subgroup. Now if $S(E\backslash\{1\})S \cap H(1) \neq \square$, then $H(1) \subseteq S(E\backslash\{1\})S$, so that $1 = pqr$ for some p, $r \in S$ and $q \in E\backslash\{1\}$. In view of 1.17, $1 = pqr$ implies that p, q, $r \in H(1)$, and in particular $q = 1$ (which is a contradiction). It follows that $S(E\backslash\{1\})S \cap H(1) = \square$.

We claim that $H(1) = \phi^{-1}(H_T(1))$. Now $\phi(H(1))$ is a group, $1 \in \phi(H(1))$, and hence $\phi(H(1)) \subseteq H_T(1)$. Since $S(E\backslash\{1\})S \cap H(1) = \square$, $\phi^{-1}(H_T(1)) \subseteq H(1)$, and we conclude that $\phi(H(1)) = H_T(1)$. Since $\phi|S\backslash S(E\backslash\{1\})S$ is injective, we have that $H(1) = \phi^{-1}(H_T(1))$.

To see that T satisfies the hypothesis of this theorem, we note that T is a compact semigroup with 1; $V_0 = E(T)\backslash\{0\}$ is an open set in V with $V_0 \cap E(T) = \{1\}$; and since ϕ is quotient, $H(1)$ is not open in S, and $\phi^{-1}(H_T(1)) = H(1)$, we have that $H_T(1)$ is not open in T.

Before proceeding to the final step of the first reduction, let $\psi = \phi|S\backslash S(E\ \{1\})S$. Then ψ is a homeomorphism and if x, y, $xy \in T\backslash\{0\}$, then $\psi^{-1}(xy) = \phi^{-1}(x)\phi^{-1}(y)$.

To complete the first reduction, we need to establish that if T contains the desired local one-parameter semigroup, then so does S. For this purpose, suppose that $\tau : [0,1] \rightarrow T$ is a local one-parameter semigroup such that $\tau(0) = 1$, and $\tau(a) \notin H_T(1)$ if $0 \leqslant a \leqslant 1$. Note that if $\tau(x) = 0$, then $x = 1$, since $\tau(x) = 0$ implies $\tau(x + (1 - x)) = 0 = \tau(x)$ and since τ is injective, $x + (1 - x) = x$, $1 - x = 0$, and $x = 1$. Now define $\sigma : [0,1] \rightarrow S$ as follows:

$$
\begin{array}{ccccccc}
 & & & \sigma & & & \\
[0,1] & \longrightarrow & [0,\tfrac{1}{2}] & \xrightarrow{\ \tau\ |\ [0,\tfrac{1}{2}]\ } & T\backslash\{0\} & \xrightarrow{\ \varphi^{-1}\ } & S \\
 & & & & & & \\
x & \longmapsto & \tfrac{1}{2}x & & & &
\end{array}
$$

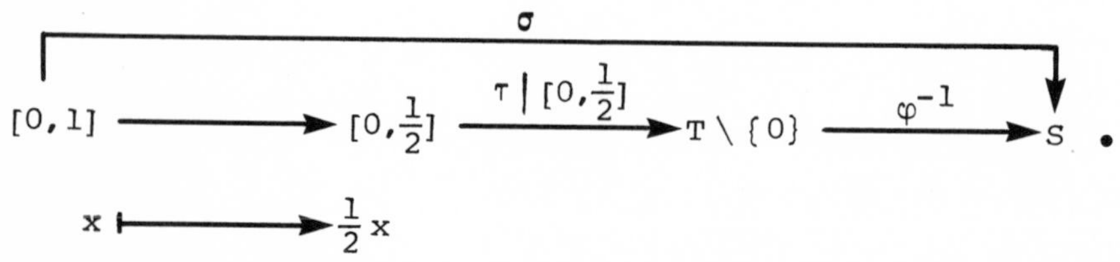

Then σ is the desired local one-parameter semigroup. This completes the first reduction and we hereafter assume that $E = \{0,1\}$.

II. We Reduce to the Case that S is an Irreducible Semigroup, E = {0,1}, and S is an H-Chain.

For $\mathcal{U}$ open in $S \times S$ containing $\Delta = \Delta_S$, set $\mathcal{V} = \{(x,y) : (x,y) \cup \Delta(x,y) \cup (x,y)\Delta \cup \Delta(x,y)\Delta \subset \mathcal{U}\}$. Then $\mathcal{V}$ is an open Δ-ideal and $\Delta \subset \mathcal{V} \subset \mathcal{U}$ by 5.7. For each $\mathcal{U}$, let U be an open set in S containing 1 such that $U \times U \subset \mathcal{V}$.

Let us first show that $U \not\subset H(1)$. Then $H(1) = \cup \{Uh : h \in H(1)\}$. For $h \in H(1)$, we claim that Uh is open. To see this, let $x \in Uh$. Then $xh^{-1} \in U$ and $Wh^{-1} \subset U$ for some open set W containing x. Thus $x \in W \subset Uh$ and Uh is open. It follows that H(1) is open and we have a contradiction. We conclude that $U \not\subset H(1)$.

Now for each $\mathcal{U}$ open in $S \times S$ containing Δ, let $x_{\mathcal{U}} \in U\backslash H(1)$.

We claim that $\theta(x_{\mathcal{U}})$ is $\mathcal{U}$-connected. To see this observe that $(1,x_{\mathcal{U}})$ and $(x_{\mathcal{U}},1) \in \mathcal{V}$, and since $\mathcal{V}$ is a Δ-ideal, $(x^k,x^{k\pm1}) \in \mathcal{U}$ for all $k \in \mathbb{N}$, where $x_{\mathcal{U}}^0 = 1$. If $x_{\mathcal{U}}^m$ and $x_{\mathcal{U}}^n \in \theta(x_{\mathcal{U}})$ for $m > n$ in $\mathbb{N}$, then $(x_{\mathcal{U}}^m,x_{\mathcal{U}}^{m-1}),\ldots,(x_{\mathcal{U}}^{n+1},x_{\mathcal{U}}^n) \in \mathcal{U}$, and hence $\theta(x_{\mathcal{U}})$ is $\mathcal{U}$-connected.

We next observe that $\theta(x)$ is an H-chain for each $x \in S\backslash H(1)$. Now $\theta(x)$ is abelian, so that $L = R = H = \mathcal{D}$ (on $\theta(x)$). If $m < n$ in $\mathbb{N}$, then $\theta(x)x^n \subset \theta(x)x^m$, so that $x^n \leq_L x^m$, and hence $\theta(x)$ is an H-chain.

For $x \in S\backslash H(1)$, it is straightforward to prove that $\Gamma(x) = \theta(x) \cup M(\Gamma(x))$, $M(\Gamma(x))$ is a group, and $\Gamma(x)$ is an H-chain.

Now let $\{A_{\mathcal{U}_\alpha}\}$ be a universal subnet of $\{\theta(x_{\mathcal{U}}) : \Delta \subset \mathcal{U}, \mathcal{U}$ open in $S \times S\}$ (see [Kelly, 1955]). We will write A_α for $A_{\mathcal{U}_\alpha}$.

We claim that $\lim\sup A_\alpha = \lim\inf A_\alpha$. Now $\lim\inf A_\alpha \subset \lim\sup A_\alpha$ is immediate. Let $x \in \lim\sup A_\alpha$, and let W be an open set containing x. We will show that $W \cap A_\alpha$ is eventually non-empty. Let $A = \{B \subset S : W \cap B \neq \Box\}$. Then since $x \in \lim\sup A_\alpha$, $\{A_\alpha\}$ is frequently in A, and since it is universal, it is eventually in A. It follows that $x \in \lim\inf A_\alpha$, and hence $\lim\sup A_\alpha = \lim\inf A_\alpha$.

Let $T = \lim A_\alpha = \lim\sup A_\alpha = \lim\inf A_\alpha$.

We claim that T is a compact abelian subsemigroup of S. That T is compact follows from the fact that lim A_α is closed. Since multiplication is continuous and each A_α is an abelian subsemigroup of S, so is T.

To see that T is an H-chain, observe that T = lim A_α = lim $\overline{A}_\alpha$, and that each $\overline{A}_\alpha$ is a compact abelian subsemigroup of S which is an H-chain.

We claim that $0 \in T$. To see this note that 0 is the idempotent in each $\Gamma(x_{\mathcal{U}})$, since $x_{\mathcal{U}} \notin H(1)$. Thus $0 \in \overline{A}_\alpha$ for each α, and hence $0 \in T$.

We claim that $1 \in T$. Let $1 \in W$, where W is an open subset of S. We want to show that $W \cap A_\alpha$ is frequently non-empty. Let D denote the directed set of the net $\{A_\alpha\}$, and let $\alpha \in D$. For each $x \neq 1$, let W_x be an open set containing x such that $1 \notin W_x$. Then $\mathcal{U} = (W \times W) \cup \bigcup\{W_x \times W_x : x \neq 1\}$ is an open subset of $S \times S$ containing Δ. Choose $\beta \in D$ such that $\beta \geqslant \alpha$ and $\mathcal{U} \subset \mathcal{U}_\beta$. This is possible since $\{A_\alpha\}$ is a subnet of $\{\theta(x_{\mathcal{U}})\}$. We show that $U_\beta \subset W$. Now $U_\beta \times U_\beta \subset \mathcal{V}_\beta \subset \mathcal{U}_\beta \subset \mathcal{U}$. Take $p \in U_\beta$. Then $(p,1) \in U_\beta \times U_\beta$, and hence $(p,1) \in \mathcal{U}$. Since $1 \notin W_x$ for all $x \neq 1$, $(p,1) \in W \times W$, and $p \in W$ and $U_\beta \subset W$. Thus $x_{\mathcal{U}_\beta} \in W$, and $W \cap A_\beta \neq \Box$ $(A_\beta = \theta(x_{\mathcal{U}_\beta}))$. It follows that $W \cap A_\alpha$ is frequently non-empty, and $1 \in \limsup A_\alpha = T$.

We next claim that T is connected. Suppose that T is not connected. Then $T = A \cup B$, where A and B are disjoint, non-empty, and compact. There exist neighborhoods of Δ_T, $\mathcal{U}_1$ and $\mathcal{U}_2$ such that $\mathcal{U}_1[A] \cap \mathcal{U}_2[B] = \Box$ and a symmetric neighborhood $\mathcal{U}$ of Δ_T such that $\mathcal{U} \circ \mathcal{U} \circ \mathcal{U} \subset \mathcal{U}_1 \cap \mathcal{U}_2$. Now lim $A_\alpha = T \subset \mathcal{U}[T] = \mathcal{U}[A] \cup \mathcal{U}[B]$, and hence

1. $A_\alpha \subset \mathcal{U}[A] \cup \mathcal{U}[B]$ eventually; and
2. $A_\alpha \cap \mathcal{U}[A] \neq \mathcal{U} \neq A_\alpha \cap \mathcal{U}[B]$ eventually.

Let $\alpha \in D$ such that 1. and 2. hold and $\mathcal{U}_\alpha \subset \mathcal{U}$, and let $a \in A_\alpha \cap \mathcal{U}[A]$ and $b \in A_\alpha \cap \mathcal{U}[B]$. Now A_α is $\mathcal{U}_\alpha$-connected, so there exists $x_0,\ldots,x_n \in A_\alpha$ such that $a = x_0$ and $(x_j,x_{j+1}) \in \mathcal{U}_\alpha$, $j = 0,1,\ldots,n-1$. Fix i so that $i + 1 = \min\{j : x_j \in \mathcal{U}[B]\}$. Then $x_0,\ldots,x_i \in \mathcal{U}[A]$, $x_{i+1} \in \mathcal{U}[B]$. There exist $a_1 \in A$ and $b_1 \in B$ such that $(a_1,x_1) \in \mathcal{U}$ and $(b_1,x_{i+1}) \in \mathcal{U}$. Thus $(x_i,x_{i+1}) \in \mathcal{U}_\alpha \subset \mathcal{U}$ implies

$(a_1,b_1) \in \mathcal{U} \circ \mathcal{U} \circ \mathcal{U} \subset \mathcal{U}_1 \cap \mathcal{U}_2$, so that $b_1 \in \mathcal{U}_1[A] \cap \mathcal{U}_2[B]$. This contradicts the fact that $\mathcal{U}_1[A] \cap \mathcal{U}_2[B] = \Box$. We conclude that T is connected.

Now T contains an irreducible subsemigroup K by 5.11, and since T is abelian, we have that K is abelian. It follows from 5.12 that K is divisible, and hence by 5.4, $H_K(1)$ is divisible. From 5.5, we see that $H_K(1)$ is connected.

It is clear that K satisfies the hypothesis of the theorem.

To see that K is an $\mathcal{H}$-chain construct an $\mathcal{H}$-chain C in K just as T was constructed from S. Since K is irreducible, K = C, and hence K is an $\mathcal{H}$-chain.

To complete the reduction we need only observe that if σ : $[0,1] \to K$ is a local one-parameter semigroup with $\sigma(0) = 1$ and $\sigma(a) \notin H_K(1)$ when $0 < a \leqslant 1$, then σ works for S.

Final Step. We hereafter assume that S be an irreducible abelian monoid such that $E = \{0,1\}$, S is an $\mathcal{H}$-chain, H(1) is not open in S, and there exists an open set V in S containing 1 such that $E \cap V = \{1\}$. Then there exists a one-parameter semigroup $\sigma : \mathbb{H} \to S$ such that $\sigma(0) = 1$ and $\sigma(a) \notin H(1)$ for $0 < a$.

We will establish the existence of a local one-parameter semigroup $\sigma : [0,1] \to S$ such that $\sigma(0) = 1$ and $\sigma(a) \notin H(1)$ for $0 < a \leqslant 1$. The result will follow from 5.6.

Now in the case that $H(1) = \{1\}$, we have that $H(x) = x$ for each $x \in S$, so that $\leqslant_{\mathcal{H}}$ is a total order on S. Since $\leqslant_{\mathcal{H}}$ is closed, S is a continuum, $0 = \inf S$, and $1 = \sup S$, we have that S is an arc with endpoints 0 and 1. In view of 4.12, we see that S is topologically isomorphic to either I_u or I_n. Now $t \to e^{-t}$ is the desired local one-parameter semigroup on I_u, and $t \to (3/4)^t$ works for I_n. Thus in the case $H(1) = \{1\}$, the conclusion of the Lemma follows.

We assume that $H(1) \neq \{1\}$, and let $\chi : H(1) \to \mathbb{R}/\mathbb{Z}$ be a nontrivial character of H(1) (5.8). We will contradict the assumption that $H(1) \neq \{1\}$, and thus establish the theorem, since the $H(1) = \{1\}$ case has been proved. Let $K = \{h \in H(1) : \chi(h) = 1\}$ denote the kernel of χ and let $\rho = \{(x,y) \in S \times S : x \in yK\}$.

We claim that ρ is a closed congruence on S. Clearly, ρ is closed. Since $1 \in K$, $x \in xK$ for each $x \in S$, so that ρ is reflexive. If $x \in yK$ and $y \in zK$, then $x \in zK$; and hence ρ is transitive. If $x \in yK$, then $x = yh$ for $h \in K$, and $y = xh^{-1}$, so that $y \in xK$, and we have that ρ is symmetric. Now if $x \in yK$ and $s \in S$, then $x = yk$, $k \in K$, and $sx = (sy)k$, and since S is abelian, ρ is a congruence on S.

Note that the ρ-class of 0 is $\{0\}$.

We claim that ϕ is open. To establish this let U be an open subset of S. Let $x \in \phi^{-1}\phi(U)$, then $\phi(x) = \phi(u)$ for some $u \in U$, so that $(x,u) \in \rho$. Now $x = uh$ for some $h \in K$ and hence $xh^{-1} = u$. By continuity of multiplication in S, there exists an open set W in W such that $x \in W$ and $Wh^{-1} \subset U$. Thus $W \subset Uh$ and so $W \subset \phi^{-1}\phi(U)$. It follows that $\phi(U)$ is open, and hence ϕ is open.

We claim that S/ρ is irreducible. For this purpose let T be an irreducible subsemigroup of S/ρ. We will show that $S/\rho = T$. Let $T' = \phi^{-1}(T)$ and let $\phi' = \phi|T'$. Then ϕ' is clearly continuous and surjective. To see that ϕ' is open, let G be open in T'. Then $G = P \cap T'$ for some open set P in S, and $\phi'(G) = \phi(P) \cap T$ (5.9(a)); which is open in T. Thus ϕ' is open. In 5.9 (b), let $X = T'$, $Y = T$, $f = \phi'$, and $A = \{0\}$. Then $\phi'^{-1}(0) = 0$, so T' is connected. Clearly, $0, 1 \in T'$, and T' is a compact subsemigroup of S. Since S is irreducible, $S = T'$, and hence $T = S/\rho$. We conclude that S/ρ is irreducible.

Now $\chi : H(1) \to \mathbb{R}/\mathbb{Z}$ is non-trivial and H(1) is connected by 5.5 and 5.12. Thus χ is surjective. It follows that H(1)/K is topologically isomorphic to $\mathbb{R}/\mathbb{Z}$ (K is the kernel of χ).

We claim that $\phi(H(1)) = H_{S/\rho}(1)$. Since $\phi(H(1))$ is a group, it is clear that $\phi(H(1)) \subset H_{S/\rho}(1)$. Let $x \in S$ such that $\phi(x) \in H_{S/\rho}(1)$. Then $\phi(x)\phi(y) = 1 = \phi(xy)$ for some $y \in S$, and $xy \in 1 : K \subset H(1)$. Thus $x, y \in H(1)$, and $H_{S/\rho}(1) \subset \phi(H(1))$. It follows that $\phi(H(1)) = H_{S/\rho}(1)$.

Observe that if $\phi(x) \in H_{S/\rho}(1)$, then $x \in H(1)$, as was shown in the preceding paragraph. It follows that $H(1) = \phi^{-1}(H_{S/\rho}(1))$.

Note that $\rho \cap H(1) \times H(1) = \{(x,y) \in H(1) \times H(1) : xy^{-1} \in K\}$ is the kernel congruence on χ. Thus, by 5.9 (c), we have that $\phi(H(1)) = H_{S/\rho}(1)$ is topologically isomorphic to H(1)/K.

We have established that the following diagram commutes:

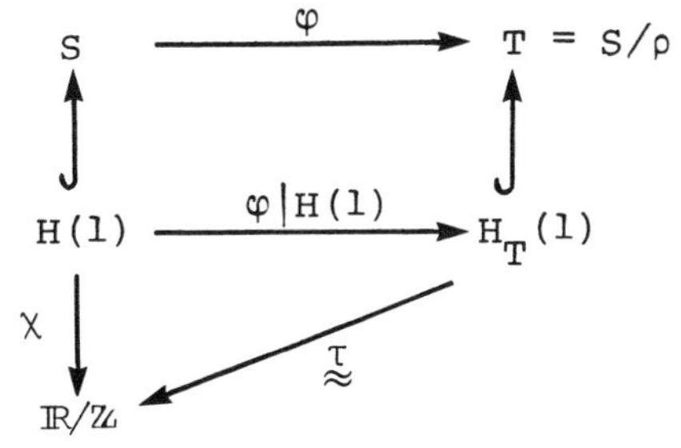

Let $T = S/\rho$. Then T is a compact abelian irreducible semigroup, $E(T) = \{0,1\}$, T is an $\mathcal{H}$-chain, and $H_T(1) = \mathbb{R}/\mathbb{Z}$. We will obtain the desired contradiction by showing that no such T exists.

We identify $H = H_T(1)$ with $\mathbb{R}/\mathbb{Z}$. For $\alpha, \beta \in \mathbb{R}$, we use $[\alpha,\beta] = \{x \in \mathbb{R}/\mathbb{Z} : \alpha \leqslant \arg x \leqslant \beta\}$. Observe that, since T is normal, there exists an open set U in T such that $\overline{U} \cap ([\frac{\pi}{2},\frac{3\pi}{2}] \cup \{0\}) = \Box$.

Let $f : T \to T$ be defined by $f(x) = x^2$.

We claim that $1 \in f(U)^0$ (interior in T). Note that $x \in A^0$ is equivalent to each of (1) $x \in T\ \overline{(T\setminus A)}$; (2) $x \notin \overline{T\setminus A}$; and (3) for each net $\{x_\alpha\} \to x$, we have $\{x_\alpha\}$ is frequently in A. To prove that $1 \in f(U)^0$, let $\{x_\alpha\}$ be a net in T with $\{x_\alpha\} \to 1$. Since T is divisible, there exists a net $\{y_\alpha\}$ such that $y_\alpha^2 = x_\alpha$ for each α and $\{y_\alpha\}$ clusters to some $y \in T$. Now $\{y^2\} = \{x_\alpha\}$ clusters to y^2 and hence $y^2 = 1$. It follows that $y = \pm 1$. Now $(-y_\alpha)^2 = x_\alpha$, where $-y_\alpha = (-1)y_\alpha$, so one of $\{y_\alpha\}$ or $\{-y_\alpha\}$ clusters to 1, and hence is frequently in U. It follows that $\{x_\alpha\}$ is frequently in $f(U)$, and hence $1 \in f(U)^0$.

Using the normality of T, there exists an open set W such that $1 \in W \subset U \cap f(U)$, $W \cap H = (-\varepsilon,\varepsilon)$, and $W \cap [2\varepsilon, 2\pi - 2\varepsilon] = \Box$ for some $\varepsilon > 0$.

Let $\mathcal{P}$ denote the family of all closed proper ideal of T, and for each $I \in \mathcal{P}$, let $W_I = W\setminus I$.

We claim that $W_I \subset f(W_I)$ for some $I \in \mathcal{P}$. Suppose this is not true. Then for each $I \in \mathcal{P}$, there exists $x_I \in W_I\setminus f(W_I)$. Now $W_I \subset W \subset f(U)$, so there exists $y_I \in U$ such that $y_I^2 = x_I$. Note that $x_I \notin I$, and hence $y_I \notin I$, since I is an ideal for each $I \in \mathcal{P}$. Thus $y_I \notin W$

(otherwise we would have $y_I \in W$, $y_I \in W_I$, and hence $x_I = y_I^2 \in f(W_I)$) for each $I \in P$. Now $\{T\backslash I : I \in P\}$ is a basis for the neighborhoods of H. Let $\{y_I\}$ cluster to y. Then $y \in \overline{U}\backslash W$. Note that $y \in H$, since otherwise if $y \notin H$, we would have disjoint open sets A and B with $H \subset A$, $y \in B$, and $T\backslash I \subset A$ for some $I \in P$. For $J \in P$ and $J \geqslant I$ (i.e. $T\backslash J \subset T\backslash I$) we would have $T\backslash J \subset A$, $y_J \in A$, and $\{y_I\}$ could not cluster to y. We conclude that $y \in H$. By considering subnets, we can assume that $\{y_I\} \to y \in H \cap (\overline{U}\backslash W)$. Then $\{x_I\} \to y^2 \in H \cap \overline{W}$. Since $y \in H \cap (\overline{U}\backslash W)$, we have $\arg y \in [\varepsilon,\pi/2) \cup (-\pi/2,-\varepsilon]$, so that $\varepsilon \leqslant |\arg y| < \pi/2$. We obtain that $2\varepsilon \leqslant |\arg y^2| < \pi$. On the other hand $y^2 \in W \cap H$ and hence $|\arg y^2| < 2\varepsilon$. This contradiction proves that $W_I \subset f(W_I)$ for some $I \in P$.

Let I be a closed proper ideal of T such that $W_I \subset f(W_I)$. Note that $W_I \not\subset H$, since H is not open in T.

Pick $x_0 \in W_I\backslash H$ and recursively define $x_n \in W_I$ such that $x_n^2 = x_{n-1}$ for all $n \in \mathbb{N}$. Note that $x_n \notin H$ for all $n \in \mathbb{N}$, since $x_0 \notin H$.

Note that T/H is an arc by the same argument given in the case $H(1) = \{1\}$. Thus, as in that case, T/H is topologically isomorphic to either I_u or I_n.

We claim that the set of cluster points of $\{x_n\}$ are contained in H. Let y be a cluster point of $\{x_n\}$. Then $\{x_{n-1}\} = \{x_n^2\}$ clusters to y^2. Inductively, we see that $\{x_n\}$ clusters to y^{2^k} for each $k \in \mathbb{N}$ (and for $k = 0$). Now let $\psi : T \to T/H$ denote the natural quotient homomorphism. Then $\{\psi(x_n)\}$ clusters to $\psi(y)^{2^k}$ for all $k \geqslant 0$. If $y \notin H$, then $\{\psi(y)^{2^k}\} \to 0$, and hence $\{\eta(x_n)\}$ clusters to 0. Let N be open in T with $0 \in N$, and $\overline{N} \cap \overline{U} = \Box$. Then there exists an open ideal J such that $0 \in J \subset V$. Note that $\psi^{-1}\psi(J) = J$, and hence $\{x_n\}$ is frequently in $J \subset T\backslash W_I$. This contradiction proves that $y \in H$.

We claim that $\{x_n\} \to 1$. From the preceding paragraph, we have that the cluster points of $\{x_n\}$ are in $\overline{W}_I \cap H \subset \overline{U} \cap H \subset (-\pi/2,\pi/2)$. If $\{x_n\}$ clusters to h, then $\{x_n\}$ clusters to h^{2^k} for all k. If $h \neq 1$, then $h^{2^k} \in [\pi/2,3\pi/2]$ for some k. Thus $h = 1$ is the unique cluster point of $\{x_n\}$.

We may assume that $\overline{W} \subset U$ and $W^2 \subset U$.

We claim that $\overline{U}$ contains no non-trivial subsemigroups. Let $z \in U\backslash\{1\}$. If $z \in H$, then some power of Z is in $[\pi/2, 3\pi/2]$ and thus not in $\overline{U}$. If $z \notin H$, then $\psi(z) \neq 1$, and hence $\{\psi(z)^n\} \to 0$. Thus $\{y^n\}$ is is eventually in some open ideal J of T with $J \subset T\backslash\overline{U}$, and hence $\{y^n\}$ is eventually in the complement of $\overline{U}$. Thus some power of each element of $\overline{U}$ is not in $\overline{U}$. We conclude that $\overline{U}$ contains no non-trivial subsemigroups.

In view of 5.10, we see that there exists an open set M in T containing 1 such that $M \subset W$ and if $x \in M$ and $n \in \mathbb{N}$, then $x^k \in U$ for all k with $1 \leqslant k \leqslant n$ and $x^n \in M$ implies that $x^k \in W$ for all k with $1 \leqslant k \leqslant n$. Now $\{x_n\} \to 1$ so there exists $m \in \mathbb{N}$ such that $x_n \in M$ when $n \geqslant m$, $n \in \mathbb{N}$. We renumber the sequence $\{x_n\}$ so that $m = 0$. Then $x_n \in M$ for all integers $n \geqslant 0$.

We claim that for all $n \geqslant 0$, $\{x_n^m : 1 \leqslant m \leqslant 2^n\} \subset W$. It will be established by induction on n. If $n = 0$, then $1 \leqslant m \leqslant 2^0 = 1$ implies $m = 1$ and $x_0^1 = x_0 \in W$. Suppose the inclusion is valid for $n - 1$, and let $1 \leqslant m \leqslant 2^n$. Then $m = 2p + r$ where $r = 0$ or 1 and p is an integer with $0 \leqslant p \leqslant 2^{n-1}$. We have $x_n^m = (x_n^2)^p x_n^r = x_{n-1}^p x_n^r \in WM \subset W^2 \subset U$. Thus, $x_n \in M$ and $x^{2^n} = x_0 \in M$, and we obtain that $x_n^m \in W$ for $1 \leqslant m \leqslant 2^n$. This proves our claim.

We claim that $\theta(x_n) \subset W \cup Tx_0$ for all n. To see this, observe that $\{x_n, \ldots, x_n^{2^n}\} \subset W$. If $m > 2^n$, then there exists $k \in \mathbb{N}$ such that $m = k + 2^n$ and $x_n^m = x_n^k x_n^{2^n} = x_n^k x_0 = Tx_0$.

We claim that $\cup\,\{\theta(x_n) : n \in \mathbb{N}\}$ is a semigroup. Now if $m < n$, then $n = m + k$ for some $k \in \mathbb{N}$ and $x_n^{2^k} = x_m$, so that $\theta(x_m) \subset \theta(x_n)$. The statement of the claim follows.

Let T_0 denote the compact semigroup $\overline{\cup\,\{\theta(x_n) : n \in \mathbb{N}\}}$, and observe that $T_0 \subset \overline{W \cup Tx_0} = \overline{W} \cup Tx_0 \subset \overline{U} \cup Tx_0$.

Now $1 \in T_0$, since $\{x_n\} \to 1$. We also have that since $x_n \notin H_T$, 0 is the idempotent in $\gamma(x_n)$, so that $0 \in T_0$. Thus T_0 is a compact semigroup with $E(T_0) = \{0,1\}$.

We claim that $H_{T_0}(1)$ is not open in T_0. Let A be open in T with $1 \in A$. Then $\{x_n\}$ is eventually in $A \backslash H_T$ implies that $\{x_n\}$ is eventually in $(A \cap T_0) \backslash H_T$, so that $A \cap T_0 \not\subset H_T \cap T_0 = H_{T_0}$ and hence H_{T_0} is not open in T_0.

Using the argument of the second reduction, there is a compact connected monoid T_1, with $0, 1 \in T_1$, and $T_1 \subset T_0$. Since T is irreducible, we have that $T_1 = T = T_0$. In particular, $H_T = H_{T_0}$.

We claim that $H_{T_0} \subset T_0 \cap H \subset (-\pi > 2, \pi/2)$. Now $x_0 \notin H_T$ and thus $Tx_0 \cap H_T = \square$. It follows that $T_0 \cap H_T \subset (\overline{U} \cup Tx_0) \cap H_T \subset \overline{U} \cap H_T \subset (-\pi/2, \pi/2)$.

We conclude that $H_T = H_{T_0} \not\subset H_T$. This contradiction yields that $H(1) = \{1\}$, and the proof of the theorem is complete. ∎

SOLENOIDAL SEMIGROUPS

We will present material in this section on solenoidal semigroups. These semigroups are the "atoms" of compact connected monoids. They are discussed in [Hofmann and Mostert, 1966].

A topological semigroup S is said to be *solenoidal* if there exists a one-parameter semigroup $\sigma : \mathbb{H} \to S$ such that $S = \overline{\sigma(\mathbb{H})}$, i.e., S contains a dense one-parameter semigroup.

5.14 Theorem. Let S be a topological semigroup and $\phi : \mathbb{H} \to S$ a continuous homomorphism. If $a < b$ in $\mathbb{H}$ and $\phi(a) = \phi(b)$, then $\phi(\mathbb{H}) = \phi([0,b])$. Hence either ϕ is injective or $\phi(\mathbb{H})$ is compact.

Proof. We present the proof in parts whose proofs parallel those of the claims in 3.1.

(i) $\phi(a + k(b - a)) = \phi(a)$ for each $k \in \mathbb{N}$;

(ii) if $s \in \mathbb{H}$ and $a < s$, then $\phi(s + k(b - a)) = \phi(s)$ for each $k \in \mathbb{N}$;

(iii) if $q \in \mathbb{H}$, $b < q$, and $q = a + k(b - a)$ for some $k \in \mathbb{N}$, then $\phi(q) = \phi(b)$; and

(iv) if $q \in \mathbb{H}$, $b < q$, and $q = a + k(b - a) + r$ for some $r \in \mathbb{H}$ with $r < b - a$, then $\phi(q) = \phi(a + r)$ and $a + r < b$. ■

5.15 Corollary. If S is a topological semigroup and $\phi : \mathbb{H} \to S$ is a continuous homomorphism, then either ϕ is constant or $\phi|[0,a]$ is injective for some $0 < a$ with $a \in \mathbb{H}$.

Proof. If $\phi|[0,a]$ is not injective for some $0 < a$, then by 5.14, there is a decreasing sequence $\{b_n\}$ in $\mathbb{H}$ with $\{b_n\} \to 0$ such that $\phi(\mathbb{H}) = \phi([0,b_n])$ for each $n \in \mathbb{N}$. Thus $\phi(\mathbb{H}) = \bigcap_{n \in \mathbb{N}} \phi([0,b_n]) = \{\phi(0)\}$, by the continuity of ϕ, and we conclude that ϕ is constant. ■

5.16 Theorem. Let G be a compact group. Then G is solenoidal if and only if there exists a continuous homomorphism $\alpha : \mathbb{R} \to G$ such that $\overline{\alpha(\mathbb{R})} = G$.

Proof. Suppose that G is solenoidal, and let $\sigma : \mathbb{H} \to G$ be a dense one-parameter semigroup. Define $\alpha : \mathbb{R} \to G$ by $\alpha(x) = \sigma(x)$ if $x \in \mathbb{H}$, and $\alpha(x) = \sigma(-x))^{-1}$ (inverse in G) if $x \in \mathbb{R}\backslash\mathbb{H}$. Then α is a continuous homomorphism and is dense, since $\alpha|\mathbb{H} = \sigma$ and σ is dense.

Suppose, on the other hand, that $\alpha : \mathbb{R} \to G$ is a dense continuous homomorphism and let $\sigma = \alpha|\mathbb{H}$. Then $\sigma : \mathbb{H} \to G$ is a continuous homomorphism. Observe that $\alpha(0)\alpha(0) = \alpha(0 + 0) = \alpha(0)$ is idempotent, and hence $\alpha(0)$ is the identity of G. Note also that $\alpha(-x)$ is the inverse of $\alpha(x)$ for each $x \in \mathbb{R}$. Since σ is a continuous homomorphism, we have that $\overline{\sigma(\mathbb{H})}$ is a compact subsemigroup of G containing the identity $\sigma(0)$ of G. Thus $\sigma(\mathbb{H})$ is a compact subgroup of G. It remains to show that σ is dense in G. For this purpose, observe that $\alpha(\mathbb{H}) \subset \overline{\sigma(\mathbb{H})}$ and the fact that $\overline{\sigma(\mathbb{H})}$ is a group imply that $\alpha((-\infty,0]) = \alpha(\mathbb{H})^{-1}$ is contained in $\sigma(\mathbb{H})$. In view of this, we see that $\alpha(\mathbb{R}) \subset \overline{\sigma(\mathbb{H})}$, and it follows that $\sigma(\mathbb{H})$ is dense in G. We conclude that G is a compact solenoidal group, and the proof of the theorem is complete. ■

5.17 Theorem. Let S be a compact solenoidal semigroup. Then S is a compact connected abelian monoid and M(S) is a compact solenoidal group.

Proof. Let $\sigma : \mathbb{H} \to S$ be a dense one-parameter semigroup. Then, since $\sigma(\mathbb{H})$ is dense in S, it is clear that S is a compact connected abelian semigroup with identity $\sigma(0)$, and hence M(S) is a compact abelian group. Let e be the identity of M(S) and define $\beta : \mathbb{H} \to M(S)$ by $\beta(x) = e\sigma(x)$ for $x \in \mathbb{H}$. Then β is a continuous homomorphism, and $\overline{\beta(\mathbb{H})} = \overline{e\sigma(\mathbb{H})} = e\overline{\sigma(\mathbb{H})} = eS = M(S)$. ■

An example that is often visualized when considering compact solenoidal semigroups is the "wind on a circle" example in Chapter 1. It is clear that $\mathbb{R}/\mathbb{Z}$ is a compact solenoidal group via the natural homomorphism $\sigma : \mathbb{R}/\mathbb{Z}$. By employing a construction due to Wallace, we will show that $\mathbb{R}/\mathbb{Z}$ can be replaced by any compact solenoidal group G to obtain a compact solenoidal semigroup which "winds down" on G.

Let G be a compact solenoidal group with $\alpha : \mathbb{H} \to G$ a dense continuous homomorphism. Let $\beta : \mathbb{H} \to \mathbb{H}^*$ be inclusion, and define $\sigma : \mathbb{H} \to G \times \mathbb{H}^*$ by $\sigma(x) = (\alpha(x), \beta(x))$ for each $x \in \mathbb{H}$. Then $S = \overline{\sigma(\mathbb{H})}$ is a compact solenoidal semigroup with $M(S) = G \times \{\infty\}$ which is topologically isomorphic to G.

Recall from the second example following 2.55 that $\hat{\mathbb{R}_d} = \beta(\mathbb{R})$ is the universal compact solenoidal group. We will use the construction above to build the universal compact solenoidal semigroup.

__*5.18 Theorem*__. Let $\$ = \{(r, b_{\mathbb{R}}(r)) : r \in \mathbb{H}\} \cup B(\mathbb{R}) \times \mathbb{H}^*$ and define $\beta : \mathbb{H} \to \$$ by $\beta(r) = (r, b_{\mathbb{R}}(r))$ for each $r \in \mathbb{H}$. Then $\beta : \mathbb{H} \to \$$ is the Bohr compactification of $\mathbb{H}$.

Proof. The proof is exactly the same as that of 3.7 with $\mathbb{M}$ replaced by $\$$, $\mathbb{N}$ replaced by $\mathbb{H}$, and $\mathbb{Z}$ replaced by $\mathbb{R}$. ■

__*5.19 Corollary*__. A compact semigroup S is solenoidal if and only if there exists a continuous surmorphism $\phi : \$ \to S$.

The semigroup $\$$, in view of 5.18 and 5.19, is called the *universal compact solenoidal semigroup*.

Solenoidal semigroups are used to construct the building blocks for irreducible semigroups called cylindrical semigroups [Hofmann and Mostert, 1966].

A compact semigroup S is said to be *cylindrical* if there exists a compact solenoidal semigroup A, a compact group G, and a continuous surmorphism $\phi : A \times G \to S$. This is clearly equivalent to S being a continuous surmorphic image of $\$ \times G$.

If $\sigma : \mathbb{H} \to S$ is a dense continuous homomorphism of $\mathbb{H}$ into a nondegenerate compact semigroup S, then according to 5.14, either σ is injective or $\sigma(\mathbb{H})$ is compact. Two ways of constructing an injective σ are via 5.17 or as in our final example below. On the other hand, if $\sigma(\mathbb{H})$ is compact, then $S = \sigma(\mathbb{H}) = \sigma([0,b])$ for some $0 < b$, since σ is not constant (5.15). It is not difficult to show that in this situation, $M(S) = \sigma([a,b])$ for some a with $0 < a \leqslant b$. If $a = b$, then of course $M(S)$ is degenerate, and if $0 < b$, then $M(S) \overset{\tau}{\approx} \mathbb{R}/\mathbb{Z}$. We present some examples to illustrate the various possible structures of $S = \sigma(\mathbb{H})$ in the latter case. Each example will be constructed by considering a closed congruence ρ on $\mathbb{H}$. Let $S = \mathbb{H}/\rho$ and let $\sigma : \mathbb{H} \to S$ denote the natural homomorphism for each .

If ρ is the closed congruence on $\mathbb{H}$ generated by $\{(x,x+1) : x \in \mathbb{H}\}$, then $S \overset{\tau}{\approx} \mathbb{R}/\mathbb{Z}$ and $\sigma(0)$ is the identity of S.

Let ρ be the closed congruence on $\mathbb{H}$ generated by $\{(x,x) : 0 \leqslant x \leqslant 1\} \cup \{(x,x+1) : 1 \leqslant x\}$. Then $M(S) \overset{\tau}{\approx} \mathbb{R}/\mathbb{Z}$ and $\sigma(1)$ is the identity of $M(S)$. The following illustration displays the structure of S:

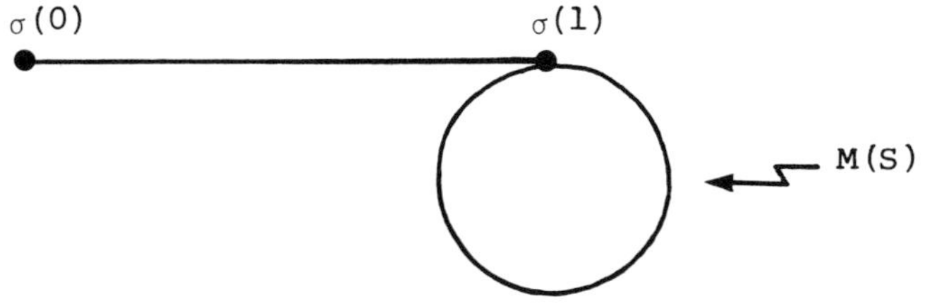

Let ρ be the closed congruence on $\mathbb{H}$ generated by $\{(x,x) : 0 \leqslant x \leqslant 1\} \cup \{(x,y) : 1 \leqslant x,\ 1 \leqslant y\}$. Then $S \overset{\tau}{\approx} \mathbb{H}/[1,\infty) \overset{\tau}{\approx} I_n$ and $M(S) = \{\sigma(1)\}$.

Let ρ be the closed congruence on $\mathbb{H}$ generated by $\{(x,x) : x \leqslant 2\} \cup \{(x,x+3) : 2 \leqslant x\}$. Then $M(S) \overset{\tau}{\approx} \mathbb{R}/\mathbb{Z}$ and $\sigma(3)$ is the identity of $M(S)$. The following illustration displays the structure of S:

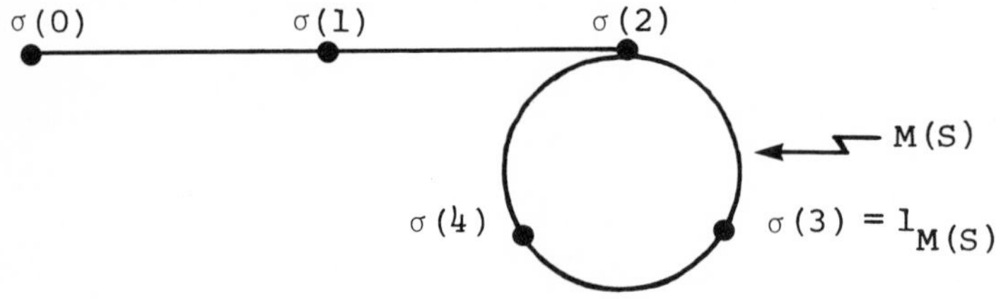

Let G be a compact solenoidal group with an injective dense continuous homomorphism $\beta : \mathbb{H} \to G$. Let $\alpha : \mathbb{H} \to \mathbb{H}/[1,\infty] = \mathbb{H}_1^*$ be the natural map, and define $\alpha : \mathbb{H} \to \mathbb{H}_1^* \times G$ by $\sigma' : (t) = (\alpha(t), \beta(t)$, $t \in \mathbb{H}$. Let $S = \overline{\sigma(\mathbb{H})}$. Then S is a compact solenoidal semigroup and σ is injective.

The final example of this chapter demonstrates that without the hypothesis in 5.13 that there are no idempotents close to 1, the conclusion need not hold. This and other interesting examples appear in [Hunter, 1961a].

Let D_n be a sequence of 2-cells (which we represent as cones in the illustration below) of diameter 1/n converging up to $\mathbb{R}^3$ to a point 1. From the center of D_n we start a one-parameter semigroup which winds down on the boundary of D_{n-1}. The multiplication on the constructed semigroup S is defined in the obvious way.

Chapter 6

COMPACT DIVISIBLE SEMIGROUPS

This chapter is devoted to the study of compact divisible semigroups in terms of local structure and in terms of global structure for some restricted classes of semigroups.

The first three sections of this chapter provide the essential tools used to establish the structure theorems in the next two sections. Expository material is contained in the final section which contains a few additional results in the area of divisible semigroups. References are provided so that the interested reader may investigate the topic deeper.

We remind the reader that some results on divisibility appear in the first section of Chapter 5.

THE SET OF DIVISIBLE ELEMENTS

Some of the basic concepts related to the set of divisible elements of a semigroup are developed in this section, along with a result which connects the notions of divisibility and unique divisibility for compact abelian semigroups.

If S is a semigroup, an element of x in S is called a [*uniquely*] *divisible element* of S provided that for each $n \in \mathbb{N}$, there exists an [unique] element $y \in S$ such that $y^n = x$. (In additive notation this reads $ny = x$.)

Observe that a semigroup S is [uniquely] divisible provided each element of S is [uniquely] divisible. Also note that each idempotent of a semigroup is a divisible element.

If S is a semigroup and $n \in \mathbb{N}$, then throughout this section we use $\phi_n : S \to S$ to denote the map $x \mapsto x^n$, and S_D to denote the set of all divisible elements of S.

Notice that if S is a semigroup, then $S_D = \cap \{\phi_n(S)\}_{n \in \mathbb{N}}$ and $E(S) \subset S_D$. Moreover, if S is abelian and $S_D \neq \square$, then S_D is a subsemigroup of S, since each ϕ_n is a homomorphism in this case. It is clear, from the continuity of multiplication, that if S is a topological semigroup, then ϕ_n is continuous for each $n \in \mathbb{N}$.

6.1 Theorem. Let S be a compact semigroup. Then:

(a) $S_D = \cap \{\phi_{n!}(S)\}_{n \in \mathbb{N}} \neq \square$;

(b) S_D is compact;

(c) $S_D \cap M(S) \neq \square$; and

(d) if S is connected, then S_D is connected.

Proof. Parts (a), (b), and (d) follow from the fact that $\{\phi_{n!}(S)\}_{n \in \mathbb{N}}$ is a descending collection, and (c) is a consequence of the fact that $E(S) \cap M(S) \neq \square$. ■

6.2 Corollary. Let S be a compact connected abelian semigroup with identity 1. Then S_D is a compact connected abelian subsemigroup of S such that $1 \in S_D$ and $S_D \cap M(S) \neq \square$.

Observe that, in fact, $M(S) \subset S_D$ for a compact semigroup S.

A semigroup S is said to be *power-cancellative* provided $a^n = b^n$ for $a, b \in S$ and $n \in \mathbb{N}$ implies that $a = b$.

Note that a divisible semigroup is uniquely divisible provided that it is power-cancellative. Also observe that a homomorphic image of a divisible semigroup is again divisible.

6.3 Theorem. Let S be a compact abelian semigroup, $S_n = S$ for each $n \in \mathbb{N}$, and define $\pi_n^m : S_m \to S_n$ by $\pi_n^m(z) = z^{m!/n!}$ for $n \leqslant m$ in $\mathbb{N}$ and $z \in S_m$. Then $\{S_n, \pi_n, \mathbb{N}\}$ is a projective system and $S_* = \varprojlim S_n$ is a compact uniquely divisible abelian semigroup. Let $\pi : S_* \to S$ denote the first projection. Then:

(a) $\pi(S_*) = S_D$;

(b) π is surjective if and only if S is divisible;

(c) π is injective if and only if S_D is power-cancellative; and

(d) π is a topological isomorphism if and only if S is uniquely divisible.

Proof. It is clear that $\pi_n^m : S_m \to S_n$ is a continuous homomorphism for each $n \leq m$ in $\mathbb{N}$. To see that $\{S_n, \pi_n^m, \mathbb{N}\}$ is a projective system, suppose that $k \leq n \leq m$ in $\mathbb{N}$ and let $z \in S_m$. Then $\pi_k^n \pi_n^m(z) = \pi_k^n(z^{m!/n!}) = (z^{m!/n!})^{n!/m!} = z^{m!/k!} = \pi_k^m(z)$, so that $\{S_n, \pi_n^m, \mathbb{N}\}$ is a projective system of compact abelian semigroups. From 2.22, we have that $S_* = \varprojlim S_n$ is a compact abelian semigroup.

To see that S_* is divisible, let $x = (x_1, x_2, \ldots) \in S_*$ and select $n \in \mathbb{N}$. Let $y = (x_n^{n!/n}, \ldots, x^{(n+k)!/(k+1)!n}, \ldots)$. Then a straightforward argument shows that $y \in S_*$ and that $y^n = x$, so that S_* is divisible.

To complete the proof that S_* is uniquely divisible, we show that S_* is power-cancellative. Suppose that $a = (a_1, a_2, \ldots)$ and $b = (b_1, b_2, \ldots)$ are in S_* and that $a^m = b^m$ for some $m \in \mathbb{N}$. Let $i \in \mathbb{N}$. We will demonstrate that $a_i = b_i$. Now $a_j^m = b_j^m$ for each $j \in \mathbb{N}$, and in particular $a_{i+m}^m = b_{i+m}^m$. Thus $a_i = a_{i+m}^{(i+m)!/i!} = b_{i+m}^{(i+m)!/i!} = b_i$, and S_* is uniquely divisible.

To prove (a), first observe that $\pi(S_*) \subset S_D$, since π is a homomorphism and S_* is divisible. For the purpose of establishing the other inclusion, let $x \in S_D$. We will construct an element $(x_1, x_2, \ldots) \in S_*$ so that $x_1 = x$. First observe that S_D is a divisible subsemigroup of S. We will construct a rooted sequence in S_D following the procedure in [Hudson, 1959]. Let $x_1 = x$ and $x_2 \in S_D$ so that $x_2^2 = x_1$. Select $x_3 \in S_D$ so that $x_3^3 = x_2$. Continuing recursively, suppose that $x_i \in S_D$ with $x_i^i = x_{i-1}$. Let $x_{i+1} \in S_D$ so that $x_{i+1}^{i+1} = x_i$. The sequence $\{x_i\}$ so constructed now has the property that $(x_1, x_2, \ldots) \in S_*$ and $\pi(x_1, x_2, \ldots) = x$, so that (a) is proved.

To prove (b), simply observe that S is divisible if and only if $S = S_D$.

To prove (c), suppose first that π is injective. We will show that S_D is power-cancellative. For this purpose suppose that $x, y \in S_D$ and $x^n = y^n$ for some $n \in \mathbb{N}$. By virtue of (a), there exist $a, b \in S_*$ such that $\pi(a) = x$ and $\pi(b) = y$. From this we obtain $\pi(a^n) = (\pi(a))^n = x^n = y^n = (\pi(b))^n = \pi(b^n)$. Since π is injective, $a^n = b^n$, and hence $a = b$ because S_* is power-cancellative. It follows that $x = \pi(a) = \pi(b) = y$ and S_D is uniquely divisible. To prove the converse, suppose that S_D is uniquely divisible, and let $u, v \in S_*$ with $\pi(u) = \pi(v)$. Denote $u = (u_1, u_2, \ldots)$ and $v = (v_1, v_2, \ldots)$. Then $u_1 = v_1$. Let $i \in \mathbb{N}$. Then u_i and v_i are in S_D and $u_i^{i!} = u_1 = v_1 = v_i^{i!}$. Since S_D is power cancellative, we have $u_i = v_i$ for each $i \in \mathbb{N}$, $u = v$, and π is injective.

Part (d) follows from (b) and (c). ■

If S is a compact abelian semigroup, then the diagram $\pi : S_* \to S$ in 6.3 is called the *uniquely divisible reflection* of S.

Our final result of this section is the categorical justification for this definition.

6.4 Theorem. Let S be a compact abelian semigroup, K a compact uniquely divisible abelian semigroup, and $\phi : K \to S$ a continuous homomorphism. Then there exists a unique continuous homomorphism $\psi : K \to S_*$ such that the diagram:

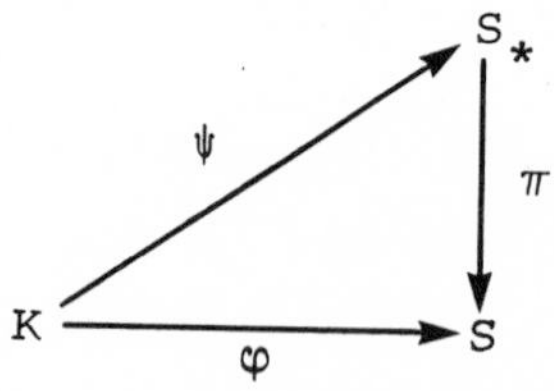

commutes.

Proof. Since K is divisible, we have that $\phi(K) \subset S_D$. In view of the unique divisibility of K, we see that for each $x \in K$, there exists a unique sequence $\{x_n\}$ in K such that $x_1 = x$ and $x_{n+1}^{n+1} = x_n$ for each $n \in \mathbb{N}$. Define $\psi : K \to S_*$ by $\psi(x) = (\phi(x_1), \phi(x_2), \ldots)$. It is clear that ψ is a well-defined homomorphism making the above diagram commute. In view of the compactness of S_*, we see that π is closed, so that ψ is continuous, since ϕ is continuous. The uniqueness of ψ is readily verified. ∎

We conclude this section with the following simple example. Others will appear later.

If $S = I_n$ (nil interval), then $S = S_D$ and $S_* \overset{\tau}{\approx} I_u$ (usual interval).

THE EXPONENTIAL FUNCTION

As we will see in a later section of this chapter, the exponential function and the associated Lie semi-algebra provides a convenient tool for working with compact divisible abelian semigroups. The definition and corresponding theory appear in [Keimel, 1967a]. We will present only a portion of Keimel's results using a somewhat different approach in that for a compact holoid S we give Hom $(\mathbb{H}^0, S)$ the compact-open topology initially and show that it is a compact uniquely divisible holoid. Keimel started by giving Hom $(\mathbb{H}^0, S)$ the topology of pointwise convergence and eventually established that this topology agrees with the compact-open topology on Hom $(\mathbb{H}^0, S)$.

An abelian semigroup S is called a *holoid* if the H-quasi-order on S is a partial order.

Observe that a holoid has degenerate H-classes and in particular, each subgroup is degenerate. Moreover, if S is an abelian semigroup, then H is a congruence and S/H is a holoid.

We begin with a special case of a more general result in [Keimel, 1967a]. This result, however, will be suitable for our purposes, and the essence of the proof of the generalization is the same.

If S is a holoid, then we use $\leqslant_S$ to denote the H-order on S.

We will use $\mathbb{H}^0$ to denote the interior of the semigroup $\mathbb{H}$, i.e., $\mathbb{H}^0$ is the additive semigroup of positive real numbers with the usual topology, and for this section, let Q denote the additive semigroup of positive rational numbers.

Observe that the semigroup $\mathbb{H}$ is a holoid as well as $\mathbb{H}^0$ and Q, and that the usual order on $\mathbb{H}$ is the reverse of the H-order. Also note that the H-order on Q is just the restriction of the H-order on $\mathbb{H}^0$ to $Q \times Q$.

Notice that if S is a holoid, then $a \leqslant_S b$ means that $a \in Sb$, since S is abelian. Also observe that if T is a holoid and $f : T \to S$ is a homomorphism, then f preserves the H-order, i.e., if $x \leqslant_T y$, then $f(x) \leqslant_S f(y)$.

6.5 Lemma. Let S be a compact semigroup and let $f : \mathbb{H} \to S$ be a homomorphism such that f is continuous at 0. Then f is continuous.

Proof. Fix $x \in \mathbb{H}$ and let $\{x_\alpha\}$ be a net in $\mathbb{H}$ converging to x. Then, in the usual order of $\mathbb{H}$, $\{x_\alpha\}$ has an ascending subnet converging to x or a descending subnet converging to x. It will be sufficient to assume that $\{x_\alpha\}$ is either descending or ascending.

Case 1. Assume that $\{x_\alpha\}$ is ascending. Then for each α, there exists $\varepsilon_\alpha \in \mathbb{H}$ such that $\{\varepsilon_\alpha\} \to 0$ and $x_\alpha = x + \varepsilon_\alpha$. We have then that $\{f(x_a)\} = \{f(x)f(\varepsilon_\alpha)\} \to f(x)f(0) = f(x + 0) = f(x)$.

Case 2. Assume that $\{x_\alpha\}$ is descending. Then for each α, there exists $\varepsilon_\alpha \in \mathbb{H}$ such that $\{\varepsilon_\alpha\} \to 0$ and $x = x_\alpha + \varepsilon_\alpha$, so that $f(x) = f(x_\alpha) = f(x_\alpha)f(\varepsilon_\alpha)$. Let y be a cluster point of $\{f(x_\alpha)\}$ in S. Then $f(x) = yf(0)$, and $y \in \overline{f(\mathbb{H})}$. Now $f(0)$ is an identity for $f(\mathbb{H})$ and hence is an identity for $\overline{f(\mathbb{H})}$. It follows that $f(x) = yf(0) = y$ and $\{f(x_\alpha)\} \to f(x)$. In any case we obtain that f is continuous. ∎

A subset U of a holoid S is said to be *order convex* provided a, $b \in U$ and $a \leqslant x \leqslant b$, $x \in S$, implies that $x \in U$. In a compact holoid S, the open order convex subsets of S form a base for the topology of S [Nachbin, 1965].

6.6 Keimel's Extension Theorem. Let S be a compact holoid and let $f : Q \to S$ be a homomorphism. Then there exists a unique continuous homomorphism $\overline{f} : \mathbb{H} \to S$ such that $\overline{f}|Q = f$.

Proof. For each $x \in \mathbb{H}$ define $\overline{f}(x) = \sup f(Q \cap (x + \mathbb{H}^0))$. To see that $\overline{f}$ is a homomorphism, let $a, b \in \mathbb{H}$. Then $\overline{f}(a)\overline{f}(b) = \sup f(Q \cap (a + \mathbb{H}^0)) \cdot \sup f(Q \cap (b + \mathbb{H}^0)) = \sup [f(Q \cap (a + \mathbb{H}^0))f(Q \cap (b + \mathbb{H}^0))] = \sup f(Q \cap (a + b) + \mathbb{H}^0) = \overline{f}(a + b)$.

In view of 6.5, in order to prove that $\overline{f}$ is continuous, we need only show that $\overline{f}$ is continuous at 0. For this purpose, let U be an open set in S such that $\overline{f}(0) \in U$. We can assume that U is order convex. Now $\overline{f}(0) = \sup f(Q)$, so that $f(a) = \overline{f}(q) \in U$ for some $q \in Q$. Since $\overline{f}$ is a homomorphism and U is order convex, we have $\overline{f}([0,q)) \subset U$, and hence $\overline{f}$ is continuous at 0.

The uniqueness of $\overline{f}$ follows from its continuity. ∎

Observe that as a consequence of 6.6, we have that each homomorphism of $\mathbb{H}^0$ into a compact holoid is continuous.

If S is a compact holoid, define $\exp : \mathrm{Hom}\,(\mathbb{H}^0,S) \to S$ by $\exp\,(f) = f(1)$ for each $f \in \mathrm{Hom}(\mathbb{H}^0,S)$. The map exp is called the *exponential function* for S.

Observe that exp is a homomorphism. We will establish its continuity as an immediate corollary of the following.

6.7 Theorem. Let S be a compact holoid and let $\pi : S_* \to S$ be the uniquely divisible reflection of S. Then there exists a topological isomorphism $\psi : S_* \to \mathrm{Hom}\,(\mathbb{H}^0,S)$ such that the diagram:

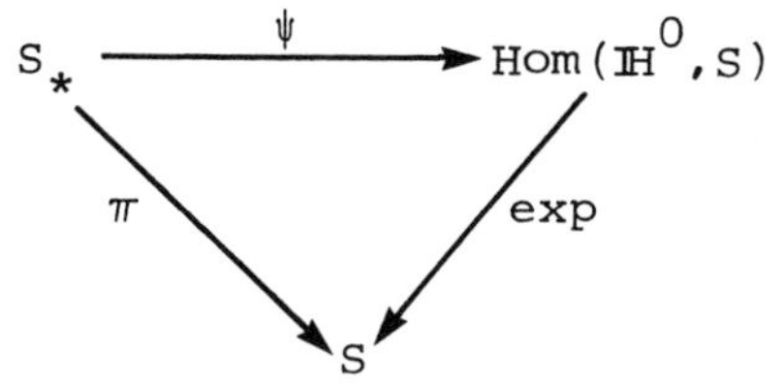

commutes.

Proof. Let $(x_1, x_2, \ldots) \in S_*$ and for m, n $\in \mathbb{N}$ define $f(\frac{m}{n}) = x_n^{m(n-1)!}$. It is straightforward to show that $f : Q \to S$ is a well-defined homomorphism, and we define $\psi(x_1, x_2, \ldots) = \overline{f}$, where $\overline{f}$ is the unique continuous extension of f to $\mathbb{H}^0$ as in 2.4. It is likewise a straightforward argument that ψ is a homomorphism and that the diagram commutes.

To see that ψ is surjective, let $g \in \text{Hom}(\mathbb{H}^0, S)$ and let $x_n = (\frac{1}{h!})$ for each $n \in \mathbb{N}$. Then $(x_1, x_2, \ldots) \in S_*$ and $\psi(x_1, x_2, \ldots) = g$ by the uniqueness of the extension in 2.4.

For the purpose of showing that ψ is injective, suppose that $\psi(x_1, x_2, \ldots) = \psi(y_1, y_2, \ldots)$. Then for each m, n $\in \mathbb{N}$, we have $x_n^{m(n-1)!} = \psi(x_1, x_2, \ldots)(\frac{m}{n}) = \psi(y_1, y_2, \ldots)(\frac{m}{n}) = y_n^{m(n-1)!}$. In particular, if we choose $k \in \mathbb{N}$, and $m = 1$ and $n = k!$, we obtain $x_k = x_{k+1}^{k+1} = x_{k!-1}^{(k!-1)(k!-2)\cdots(k+1)} = x_{k!}^{(k!-1)\cdots k!} = x_{k!}^{(k!-1)!} = y_{k!}^{(k=-1)!} = y_k$. Hence ψ is injective.

Now, since S_* is compact and $\text{Hom}(\mathbb{H}^0, S)$ is Hausdorff, it remains only to show that ψ is continuous. For this purpose suppose that K is a compact subset of $\mathbb{H}^0$, U is an open subset of S, and that $(x_1, x_2, \ldots) \in \psi^{-1}(N(K,U))$. We can assume that U is order convex. Let $g = \psi(x_1, x_2, \ldots)$. Then $g(K) \subset U$. Let W be an open set in $\mathbb{H}^0$ such that $K \subset W$ and $g(W) \subset U$. Select m, n, p, q $\in \mathbb{N}$ such that $\frac{p}{q}$, $\frac{m}{n} \in W$, and $t \in K$ implies that $\frac{p}{q} \leqslant_{\mathbb{H}^0} t \leqslant_{\mathbb{H}^0} \frac{m}{n}$. Now $g(\frac{m}{n}) = x_n^{m(n-1)!} \in U$ and $g(\frac{p}{q}) = x_q^{p(q-1)!} \in U$. Let V_n and V_q be open sets in S such that $V_n^{m(n-1)!} \cup V_q^{p(q-1)!} \subset U$, $x_n \in V_n$, and $x_q \in V_q$. Let N be the basic open set in S_* whose n^{th} factor is V_n and whose q^{th} factor is V_q with all of the other factors being copies of S. Then clearly $(x_1, x_2, \ldots) \in N$. We will demonstrate that $\psi(N) \subset N(K,U)$. For this purpose, let $(y_1, y_2, \ldots) \in N$, and let $f = \psi(y_1, y_2), \ldots)$. Then $y_n \in V_n$ and $y_q \in V_q$, so that $f(\frac{m}{n}) = y_n^{m!(n-1)!} \subset U$, and likewise $f(\frac{p}{q}) \in U$. Now let $t \in K$. Then $\frac{p}{q} \leqslant_{\mathbb{H}^0} t \leqslant_{\mathbb{H}^0} \frac{m}{n}$, so that $f(\frac{p}{q}) \leqslant_S f(\frac{m}{n})$, and since U is order convex, we have $f(t) \in U$. It follows that $f(K) \subset U$, $f \in N(K,U)$, and ψ is continuous. ∎

If S is a compact holoid, then we generally use the multiplicative notation for S and the additive notation in Hom $(\mathbb{H}^0, S)$, i.e., if f, $g \in$ Hom $(\mathbb{H}^0, S)$, then f + g is the element of Hom $(\mathbb{H}^0, S)$ such that $(f + g)(x) = f(x)g(x)$ for each $x \in \mathbb{H}^0$.

Observe that for a compact holoid S, we have $f \leqslant_{\text{Hom}\,(\mathbb{H}^0,S)} g$ if and only if $f(x) \leqslant_S g(x)$ for each $x \in \mathbb{H}^0$. With this observation, we see that Hom $(\mathbb{H}^0, S)$ is also a holoid. As a consequence of 6.7 we obtain:

6.8 Corollary. Let S be a compact holoid. Then

(a) Hom $(\mathbb{H}^0, S)$ is a compact uniquely divisible holoid; and

(b) exp : Hom $(\mathbb{H}^0, S) \to S$ is a continuous homomorphism.

If S is a compact holoid, then the semigroup Hom $(\mathbb{H}^0, S)$ has been referred to as the "Lie semi-algebra of S" (see [Hofmann and Mostert, 1969]). This term is justified by the fact that Hom $(\mathbb{H}^0, S)$ admits a scalar multiplication by elements of $\mathbb{H}$.

6.9 Theorem. Let S be a compact holoid. Then the function $\mathbb{H} \times$ Hom $(\mathbb{H}^0, S) \to$ Hom $(\mathbb{H}^0, S)$ defined by $(r,f) \mapsto rf$, where $rf(t) = f(rt)$ for each $t \in \mathbb{H}^0$, satisvies the following conditions:

(a) $1f = f$;

(b) $(rs)f = r(sf)$;

(c) $(r + s)f = rf + sf$; and

(d) $r(f + g) = rf + rg$

for all f, $g \in$ Hom $(\mathbb{H}^0, S)$ and all r, $s \in \mathbb{H}$. Moreover, the restriction of $(r,f) \mapsto rf$ to $\mathbb{H}^0 \times$ Hom $(\mathbb{H}^0, S)$ is continuous.

Proof. Conditions (a), (b), (c), and (d) are easily verified. To see that the restriction to $\mathbb{H}^0 \times$ Hom $(\mathbb{H}^0, S)$ is continuous, let $\{(r_\alpha, f_\alpha)\}$ be a net in $\mathbb{H}^0 \times$ Hom $(\mathbb{H}^0, S)$ converging to (r,f). Then for each net $\{x_\beta\} \to x$ in $\mathbb{H}^0$, we have $r_\alpha f_\alpha(x_\beta) = f_\alpha(r_\alpha x_\beta) \to f(rx) - rf(x)$, since $r_\alpha x_\beta \to rx$, and hence $r_\alpha f_\alpha \to rf$. ∎

We will employ the following result in the next section of this chapter:

6.10 Theorem. Let S be a compact holoid and let $f \in \text{Hom}(\mathbb{H}^0, S)$. Define $f^* : \mathbb{H} \to \text{Hom}(\mathbb{H}^0, S)$ by $f^*(r) = rf$. Then f^* is continuous.

Proof. Let $\{r_\alpha\} \to r$ in $\mathbb{H}$ and let $\{x_\beta\} \to x$ in $\mathbb{H}^0$. Then $\{f^*(r_\alpha)(x_\beta)\} = \{r_\alpha f(x_\beta)\} = \{f(r_\alpha x_\beta)\} = \{\overline{f}(r_\alpha x_\beta)\} \to \overline{f}(rx) = r\overline{f}(x) = rf(x) = f^*(r)(x)$, since $\{r_\alpha x_\beta\} \to rx$ in $\mathbb{H}$. It follows that $\{f^*(r_\alpha)\} \to f^*(r)$, and f^* is continuous. ∎

In [Keimel, 1967a] it is established that the functions f^* as in 6.10 are precisely the set of one parameter semigroups in $\text{Hom}(\mathbb{H}^0, S)$. However, 6.10 will suffice for our purposes.

CONE SEMIGROUPS

The primary purpose of this section is to present results which will be applied to the central theorem in the section of finite dimensional semigroups of this chapter. Results in this section appear in [Keimel, 1967a] and in [Lawson and Madison, 1971]. The reader is referred to [Keimel, 1967b] for additional material on cone semigroups.

A cone semigroup in a real vector space V is a subset C of V such that $C + C \subset C$, $C \cap (-C) = \{0\}$ and $rC \subset C$ for all $r \in \mathbb{H}$. If C is endowed with a Hausdorff topology such that (C,+) is a topological semigroup and scalar multiplication $\mathbb{H} \times C \to C$ is continuous, then C is called a *topological cone semigroup*. If $\leqslant_c$ is closed in $C \times C$, then C is called an *ordered topological cone semigroup*.

If S is a compact holoid and $e \in E(S)$ then the e-*component* of S is defined to be $C(e) = \{f \in \text{Hom}(\mathbb{H}^0, S) : \overline{f}(0) = e\}$, where $\overline{f} : H \to S$ is the unique continuous extension of f in 6.6.

Notice that if S is a compact holoid and $e \in E(S)$, then $\exp C(e) = \{x \in S : g(0) = e \text{ and } g(1) = x \text{ for some homomorphism } g : \mathbb{H} \to S\}$.

6.11 Theorem. Let S be a compact holoid and let $e \in E(S)$. Then C(e) is a cancellative holoid with identity and a scalar multiplication $\mathbb{H} \times C(e)$ defined by $(r,f) \to rf$. The element $f_e \in C(e)$ defined by $f_e(x) = e$ for all $x \in \mathbb{H}^0$ is the identity of C(e).

Proof. We want to establish that for a compact holoid S and for $e \in E(S)$, the holoid C(e) is an ordered topological cone semigroup. The following lemma combined with 6.11 will be an ingredient in the proof:

6.12 Lemma. Let S be a cancellative abelian semigroup with a scalar multiplication $\mathbb{H} \times S \to S$. Then there exists a real vector space V and an algebraic embedding $\phi : S \to (V,+)$ such that $\phi(rx) = r\phi(x)$ for each $r \in \mathbb{H}$ and $x \in S$, i.e., ϕ is a linear embedding.

Proof. It will be convenient to use additive notation for S in this proof. Let $R = \{((a,b),(c,d)) \in (S \times S) \times (S \times S) : a + d = b + c\}$. It is easily verified that R is a congruence on $S \times S$. Let V be the quotient semigroup $(S \times S)/R$ and let $\pi : S \times S \to V$ denote the natural map. A straightforward argument works to show that (V,+) is an abelian group with $\pi(a,a)$ for an identity for each $a \in S$, and $\pi(b,a)$ as the inverse of $\pi(a,b)$ for $a, b \in S$. Define $r\pi(a,b) = \pi(ra,rb)$ and $-r\pi(a,b) = \pi(rb,ra)$ for $r \in \mathbb{H}$. It is easily verified that this scalar multiplication makes V into a vector space over $\mathbb{R}$. Define $\phi : S \to V$ by $\phi(a) = \pi(2a,a)$. Then ϕ is the desired linear embedding. ∎

6.13 Lemma. Let C be a topological semigroup and let $(r,x) \mapsto rx$ be a continuous scalar multiplication from $\mathbb{H}^0 \times C$ into C. If U is open in C and $r \in \mathbb{H}^0$, then rU is open in C.

Proof. Let $x \in rU$. Then $\frac{1}{r}x \in U$, and there exists an open set V such that $x \in V$ and $\frac{1}{r}V \subset U$. We obtain that $x \in V \subset rU$ and hence rU is open. ∎

6.14 Theorem. Let S be a compact holoid and let $e \in E(S)$. Then $C(e)$ is a locally order convex ordered topological cone semigroup.

Proof. That $C(e)$ is a cone semigroup follows from 6.11 and 6.12. With the relative topology of Hom $(\mathbb{H}^0,S)$, we have that $C(e)$ is a topological semigroup, $\leqslant_c$ is closed and scalar multiplication is continuous on $\mathbb{H}^0 \times C(e) \to C(e)$ (2.8). Since Hom $(\mathbb{H}^0,S)$ is a compact holoid it is locally order convex and hence $C(e)$ is locally order convex. It remains to show that scalar multiplication is continuous at $(0,f)$ for each $f \in C(e)$. For this purpose, let $f \in C(e)$ and let $\{(r_\alpha,f_\alpha)\} \to (0,f)$ in $\mathbb{H} \times C(e)$. Let U be an open order convex neighborhood of $f_e = 0f$ in $C(e)$. Now the function $f^* : \mathbb{H} \to C(e)$ defined by $f^*(s) = sf$ is continuous (6.10), and $f^*(0) \in U$. Thus there exists $r \in \mathbb{H}^0$ such that $f^*(r) \in U$, i.e., $rf \in U$, so that $f \in \frac{1}{r}U$, which is open in $C(e)$ by 6.13. It follows that $\{f_\alpha\}$ is eventually in $\frac{1}{r}U$. Now $\{r_\alpha\} \to 0$, so that $\{r_\alpha/r\}$ is eventually below 1 in the usual order of $\mathbb{H}$, and hence $\{r_\alpha \frac{1}{r}U\}$ is eventually contained in U. We obtain that $\{r_\alpha f_\alpha\}$ is eventually in U, and hence $\{r_\alpha f_\alpha\} \to f_e = 0f$. ∎

6.15 Lemma. Let C be an ordered topological cone semigroup such that 0 has a compact neighborhood. Then C is locally compact.

Proof. Let W be an open set such that $0 \in W \subset \overline{W}$ and $\overline{W}$ is compact. Now $\{\frac{1}{n}\} \to 0$ in $\mathbb{H}$, so that for $x \in C$, we have $\{\frac{1}{n}x\} \to 0x = 0$. It follows that $\frac{1}{n}x \in W$ for some $n \in \mathbb{N}$ and hence $x \in nW \subset \overline{nW} = n\overline{W}$, and nW is open (6.13) and $n\overline{W}$ is compact. ∎

6.16 Theorem. Let S be a compact holoid and let $e \in E(S)$ such that e is isolated in $Se \cap E(S)$. Then $C(e)$ is locally compact.

Proof. In view of 6.15, we need only show that f_e has a compact neighborhood in $C(e)$. It is easily verified that f_e is isolated in the set of idempotents of f_e + Hom $(\mathbb{H}^0,S)$. Let U be an open order convex subset of Hom $(\mathbb{H}^0,S)$ containing f_e and no other idempotent of

Hom $(\mathbb{H}^0,S)$ and let V be an open set in Hom $(\mathbb{H}^0,S)$ such that $f_e \in V \subset \overline{V} \subset U$ and $\overline{V}$ is compact. We will show that $\overline{V} \cap C(e)$ is compact. For this purpose let $\{g_\alpha\}$ be a net in $\overline{V} \cap C(e)$ with $\{g_\alpha\} \to g$. Since f_e is an identity for C(e), we have that $g_\alpha \leqslant f_e$ for each α, where $\leqslant$ is the H-order on Hom $(\mathbb{H}^0,S)$. Since $\leqslant$ is closed and $g = \lim g_\alpha$, we have $g \leqslant f_e$, so that $\overline{g}(0) \leqslant_S e$. Let $e' <_S e$; i.e., $e' \neq e$. Then $g \leqslant f_{e'} < f$ and $g, f_e \in U$, so that $f_{e'} \in U$ since U is order convex. This contradicts the fact that U contains no other idempotent of Hom $(\mathbb{H}^0,S)$ except f_e. We obtain that $\overline{g}(0) = e$ and hence $g \in C(e)$. It follows that C(e) is locally compact. ■

If C is a locally compact ordered topological cone semigroup and $C^* = C \cup \{\infty\}$ is the one-point compactification of C, then the addition on C can be extended to C^* by defining $x + \infty = \infty + x = \infty$ for all $x \in C^*$, and C^* is a topological semigroup (see [Keimel, 1967b]).

Results in this section prior to this point have been those of [Keimel, 1967a]. We turn now to a useful result found in [Lawson and Madison, 1971]. The following lemma will be employed (see [Lawson and Madison, 1974]):

6.17 Lemma. Let $f : X \to Y$ be a quotient map and let Z be a locally compact space. Then $1_Z \times f : Z \times X \to Z \times Y$ is a quotient map.

Proof. Since 1_Z and f are continuous, so is $1_Z \times f$. Let U be a subset of $Z \times Y$ such that $(1_Z \times f)^{-1}(U) = W$ is open. We will show that U is open. For this purpose let $(z,y) \in U$ and let $x \in f^{-1}(y)$. Then $(z,x) \in W$ and there exist open sets P and Q such that $(z,x) \in P \times Q \subset W$. Let B be a compact neighborhood of z such that $B \subset P$ and let $A = \{p \in X : B \times \{p\} \subset W\}$. Then clearly $x \in A$, and by the compactness of B, we have that A is open.

Suppose that $p \in A$ and $q \in X$ such that $f(p) = f(q)$. Then, since $B \times \{p\} \subset W$, we have $B \times \{f(p)\} \subset U$ and hence $B \times \{f(q)\} \subset U$. It follows that $B \times \{q\} \subset W$, so that A is f-saturated. Thus f(A) is open in Y, and $f(A) \times B$ is a neighborhood of (z,y) contained in U. We conclude that U is open. ■

6.18 Theorem. Let C be a locally compact ordered topological cone semigroup. Then there exist a linear homeomorphic embedding of C into a topological vector space over $\mathbb{R}$.

Proof. Let $R = \{((a,b),(c,d)) \in (C \times C) \times (C \times C) : a + d = b + c\}$, $V = (C \times C)/R$, and $\pi : (C \times C) \to V$ be the natural map as in 6.12. Let U be a compact neighborhood of 0 in C. Then $\{nU : n \in \mathbb{N}\}$ is a cover of C by compact sets, so that $C \times C$ is locally compact and σ-compact. Observing that R is a closed congruence on $C \times C$, and giving V the quotient topology of π, we have that $(V,+)$ is a topological semigroup (1.56) and is algebraically a group by 6.12. To see that inversion is continuous in V define $\alpha : C \times C \to C \times C$ by $\alpha(x,y) = (y,x)$ and let $\theta : (V,+) \to (V,+)$ be inversion. Then clearly α is continuous and the diagram:

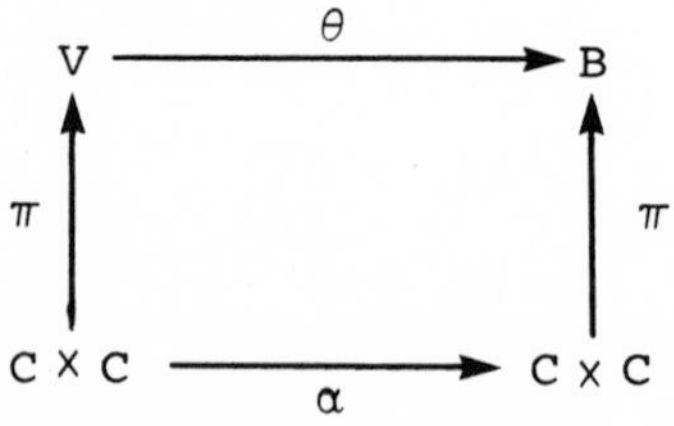

commutes. The continuity of θ now follows from the fact that π is quotient and 1.47 (2). Consequently, we have that $(V,+)$ is a topological group. We define, as in 6.12, $\phi : C \to V$ by $\phi(a) = \pi(2a,a)$ and observe that ϕ is a continuous isomorphism.

To demonstrate that ϕ^{-1} is continuous, first observe that π is quotient, $\pi^{-1}\phi(C) = \leqslant_C$, and $\leqslant_C$ is closed in $C \times C$, so that $\pi_1 = \pi|\leqslant_C$ is a quotient map. Define $\sigma : \leqslant_C \to C$ by $\sigma(x,y) = x - y$. Then the diagram:

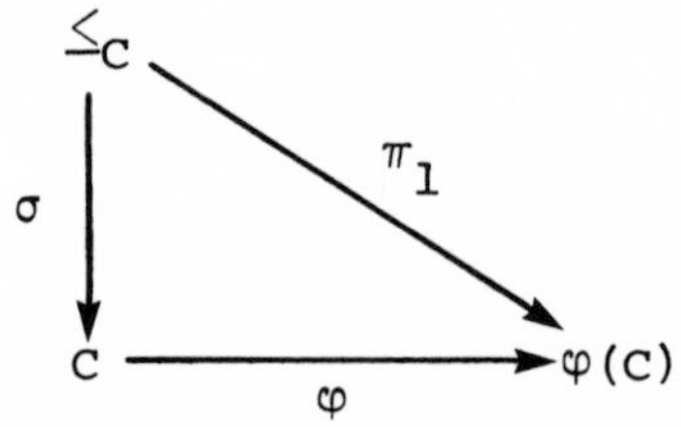

commutes. To show that ϕ^{-1} is continuous, it is sufficient to show that σ is continuous, since π_1 is quotient. For this purpose, let $\{(x_\alpha, y_\alpha)\} \to (x,y)$ in $\leq_C$. Then for each α, there exists $z_\alpha \in C$ such that $x_\alpha = y_\alpha + z_\alpha$. Now $\{z_\alpha\}$ clusters to some z in C^*, and hence $\{y_\alpha + z_\alpha\}$ clusters to y + z and converges to x. It follows that x = y + z, z = x - y, $\{\sigma(x_\alpha, y_\alpha)\} \to \sigma(x,y)$, and ϕ^{-1} is continuous.

It remains to extend the scalar multiplication on $\phi(C)$ to a continuous scalar multiplication on V so that ϕ preserves scalar products. We will use the scalar multiplication defined in the proof of 6.12 and verify that it is continuous. For this purpose we use $-\mathbb{H}$ to denote the non-positive reals. Since scalar multiplication is continuous, coordinatewise scalar multiplication is continuous on $C \times C$. With horizontal maps as scalar multiplication, we have that the diagram:

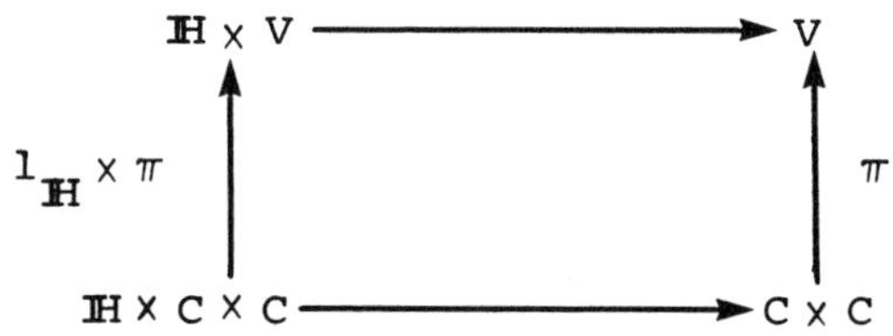

commutes. In view of 6.17, we see that $1_{\mathbb{H}} \times \pi$ is quotient, and hence the scalar multiplication $\mathbb{H} \times V \to V$ is continuous by 1.4 (2). Scalar multiplication on $-\mathbb{H} \times V \to V$ is defined to be the composition of the two functions $(r,v) \mapsto (-r,-v) \mapsto (-r)(-v) = rv$, and hence is continuous. Since scalar multiplication is continuous on the closed sets $\mathbb{H} \times V$ and $-\mathbb{H} \times V$, it is continuous on $\mathbb{R} \times V$. ∎

SUBUNITHETIC SEMIGROUPS

Here we investigate the structure of the fundamental building blocks of compact divisible semigroups. This structure was determined in [Hildebrant, 1969] with earlier contributions appearing in [Hudson, 1959] and in [Hofmann, 1960].

Throughout this section we use Q to denote the discrete additive group of rational numbers and Q^+ to denote the subsemigroup of positive rational numbers.

A topological semigroup S is said to be *subunithetic* [*unithetic*] if there exists a dense homomorphism $\phi : Q^+ \to S$ [and S is power-cancellative]. The element $\phi(1)$ is called a *subunithetic* [*unithetic*] generator of S.

Observe that a subunithetic semigroup is abelian, since it contains a dense homomorphic image of the abelian semigroup Q^+. Likewise notice that a compact subunithetic [unithetic] semigroup is divisible [uniquely divisible]. Indeed, a compact subunithetic semigroup is unithetic if and only if it is uniquely divisible.

The next result reveals a useful connection between the notions of unithetic and subunithetic.

6.19 Theorem. Let S be a compact divisible abelian semigroup and let $\pi : S_* \to S$ denote the uniquely divisible reflection of S. Then S is subunithetic if and only if S_* is unithetic.

Proof. If S_* is unithetic, then the fact that S is subunithetic is immediate from the fact that $\pi : S_* \to S$ is a surmorphism.

Suppose that S is subunithetic. Then S_* is uniquely divisible, so that it remains to show that S_* is also subunithetic. Let $\phi : Q^+ \to S$ be a dense homomorphism and define $\psi : Q^+ \to S_*$ by $\psi(r) = (\phi(r), \phi(\frac{r}{2!}), \ldots, \phi(\frac{r}{n!}), \ldots)$ for each $r \in Q^+$. It is transparent that ψ is a dense homomorphism, and hence S_I is subunithetic. ■

6.20 Theorem. Let G be a compact group. Then these are equivalent:

(a) G is subunithetic;

(b) There exists a dense homomorphism $\psi : Q \to G$; and

(c) G is solenoidal.

Proof. To prove that (a) implies (b), suppose that G is subunithetic and let $\phi : Q^+ \to G$ be a dense homomorphism. Define $\psi : Q \to G$ by $\psi(q) = \phi(q)$, $\psi(-q) = -\phi(q)$ if $q \in Q^+$ and $\psi(0) = e$ (the identity of G). It is easily verified that ψ is a dense homomorphism.

To see that (b) implies (c), let $\psi : Q \to G$ be a dense homomorphism, and let $\beta : Q \to B(Q)$ denote the Bohr compactification of Q.

Then there exists a continuous surmorphism $B(\psi) : B(Q) \to G$ such that the diagram:

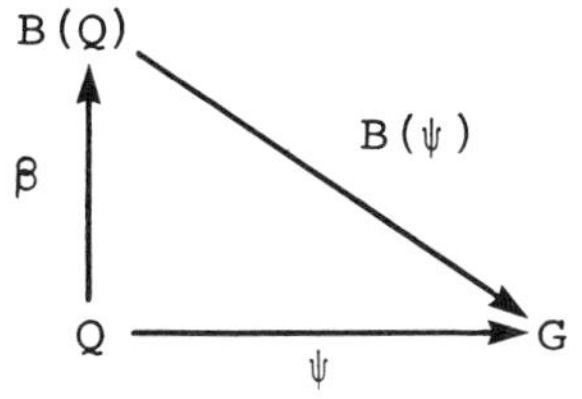

commutes. We will complete the proof of (c) by showing that B(Q) is solenoidal (in fact, B(Q) is the universal compact solenoidal group). For this proof we rely on duality and some results found in other sources. First we have that $Q^\wedge \overset{\tau}{\approx} \Sigma_a$, which is the a-adic solenoid with $a = (2,3,4,\ldots)$ or $(\mathbb{R}/\mathbb{Z})_*$ (see 25.4 and 25.5 of [Hewitt and Ross, 1963]). The group Σ_a is algebraically isomorphic to the additive discrete group of real numbers $\mathbb{R}_a$ (see Theorem 4 of [Kaplansky, 1954]), so that $(Q^\wedge)_d \approx \mathbb{R}_d$, and hence $B(Q) \overset{\tau}{\approx} (\mathbb{R}_d)^\wedge$, which is the universal compact solenoidal group (see [Hofmann and Mostert, 1966]).

That (c) implies (a) is transparent. ■

It is interesting to note that B(Q) $(\overset{\tau}{\approx} (\mathbb{R}_d)^\wedge)$ can also be expressed as Σ_a^c (the cartesian product of c copies of Σ_a with $a = (2,3,\ldots)$). This is discussed in 25.5 of [Hewitt and Ross, 1963].

We turn now to the study of the structure of compact subunithetic semigroups which are not groups.

Suppose that S is a uniquely divisible abelian semigroup, $x \in S$, and $r \in Q^+$ with $r = m/n$ for $m, n \in \mathbb{N}$ with m and n relatively prime. Then we can define $x^r = (x^{1/n})m$, where $x^{1/n}$ is the unique n^{th} root of x. Moreover, it is easily verified that x^r is independent for the choice of m and n, i.e., m and n need not be relatively prime. Observe that if S is a unithetic semigroup and $\phi : Q^+ \to S$ is a dense homomorphism, then $S = \{x^r : r \in Q^+\}^-$, where $\phi(1) = x$ is a unithetic generator.

If S is a subunithetic semigroup and $\phi : Q^+ \to S$ is a dense homomorphism, we let $N(r) = \{\phi(s) : s < r\}^-$ and $M(r) = \{\phi(s) : r < s\}^-$ for each $r \in Q^+$.

6.21 Theorem. Let S be a compact unithetic semigroup which is not a group, and let $\phi : Q^+ \to S$ be a dense homomorphism. Then:

(a) S has an identity 1 and $H(1) = \cap \{N(r)\}_{r \in Q^+}$;

(b) $M(S) = \cap \{M(r)\}_{r \in Q^+}$ (Let e be the identity of M(S));

(c) $E(S) = \{1, e\}$;

(d) There exists a one-parameter semigroup $\sigma : \mathbb{H} \to S$ such that $S = \overline{\sigma(\mathbb{H})}H(1)$;

(e) S is connected; and

(f) M(S) and H(1) are solenoidal groups.

Proof. Observe that M(r) is a closed ideal of S for each $r \in Q^+$. To prove (a), let $N = \cap \{N(r)\}_{r \in Q^+}$. Then N is a non-empty compact subset of S, since it is the intersection of a descending family of non-empty compact subsets of S.

To see that N is a subsemigroup of S, let a, b ∈ N and let W be an open subset of S containing ab. Then there exist open subsets U and V in S such that a ∈ U, b ∈ V, and UV ⊂ W. Let $r \in Q^+$ and let $r_1, r_2 \in Q^+$ such that $r_1, r_2 < r/2$, $\phi(r_1) \in U$, and $\phi(r_2) \in V$. Then $\phi(r_1)\phi(r_2) = \phi(r_1 + r_2) \in W$ and $r_1 + r_2 < r$. It follows that ab ∈ N and N is a compact subsemigroup of S.

Let f be an idempotent in N, $r \in Q^+$, and suppose, for the purpose of contradiction, that $fz \neq \phi(r)$ for all z ∈ S. Then, using continuity of multiplication and the compactness of S, we obtain open sets U and V in S with f ∈ U, $\phi(r) \in V$ and US ∩ V = □. Now let $s \in Q^+$ such that $s < r$ and $\phi(s) \in U$. Thus $\phi(r) = \phi(s + r - s) = \phi(s)\phi(r - s)$ is in US ∩ V, which is a contradiction. It follows that $fz = \phi(r)$ for some z ∈ S, so that $f\phi(r) = f(fz) = f^2z = fz = \phi(r)$, and f is an identity for $\phi(Q^+)$ and hence is an identity for S. Since f was chosen just to be an idempotent of N and is thus an ideneity for N, we have that N is a compact group. We denote this identity henceforth by 1. To complete the proof of (a), we need to verify that N = H(1). It is immediate that N ⊂ H(1).

Suppose, by way of contradiction, that $H(1)\backslash N \neq \square$ and let $y \in H(1)\backslash N$. Then there exists an open set of U such that $y \in U$ and $U \cap N = \square$. We have that $U \cap N(r) = \square$ for some $r \in Q^+$ and hence $U \subset M(r)$. Since M(r) is an ideal of S, we have that $M(r) \subset S\backslash H(1)$ by 1.34 and hence $y \in S\backslash H(1)$. This contradiction yields that $N = H(1)$ and the proof of (a) is complete.

To prove (b), let $P = \cap \{M(r)\}_{r \in Q^+}$. Then P is a closed ideal of S and hence $M(S) \subset P$. For the purpose of establishing (b), it will suffice to demonstrate that P is a group. Let $x \in P$. Then clearly $xP \subset P$. Suppose, for the purpose of contradiction, that $y \notin xP$ for some $y \in P$. Let U, V, and W be open subsets of S such that $y \in U$, $x \in V$, $P \subset W$ and $U \cap VW = \square$, and let $r \in Q^+$ such that $\phi(r) \in V$. From the definition of P, there exists $s \in Q^+$ such that $r < s$, $\phi(s) \in U$ and $\phi(s - r) \in W$. Thus $\phi(s) = \phi(r)\phi(s - r)$ is in $U \cap VW$, contradicting $U \cap VW = \square$. We obtain that $xP = P$, P is a group, $M(S) = P$, and the proof of (b) is complete.

For the purpose of establishing (c), let $f \in E(S) \cap (S\backslash M(S))$. Then $f \in N(s)$ for some $s \in Q^+$. Choose $r \in Q^+$ and let U be an open subset of S containing f. We will prove that $f \in H(1)$ (and hence $f = 1$) by showing that $U \cap N(r) \neq \square$. Let $n \; \mathbb{N}$ such that $s/n < r$. Since S is unithetic, the map $\alpha : S \to S$ defined by $\alpha(x) = x^{1/n}$ for $x \in S$ is a homeomorphism of S onto itself and clearly $\alpha(f) = f$. Let V be an open subset of S such that $f \in V$ and $\alpha(V) \subset U$, and let $p \in Q^+$ such that $p < S$ and $\phi(p) \in V$ (here we use that $f \in N(s)$). Then $\alpha\phi(p) \in U$ and again it is clear that $\alpha\phi(p) = \phi(\frac{p}{n})$. We now have that $\frac{p}{n} < r$ and $\phi(\frac{p}{n}) \in U$, so that $U \cap N(r) \neq \square$, and the proof of (c) is complete.

To prove (d), we first show that H(1) is not open in S. Otherwise we would have $\phi(r) \in H(1)$ for some $r \in Q^+$ and hence $\phi(ar) = \phi(r)^2 \in H(1)$, so that $H(1) \cap M(r) \neq \square$; contradicting 1.34. We conclude that H(1) is not open in S. Combining this fact with (c), we have a one-parameter semigroup $\sigma : \mathbb{H} \to S$ such that $\sigma(0) = 1$ and $\sigma(t) \notin H(1)$ for $0 < t$. Now let $T = \overline{\sigma(\mathbb{H})}H(1)$ and observe that H(1) is connected, since it is divisible. Then T is the continuous homomorphic image of the compact connected semigroup $\overline{\sigma(\mathbb{H})} \times H(1)$ under $(a,b) \to ab$. It follows that T is connected. Using that 1, $e \in T$ and that for each $x \in S$, we have that Tx is a connected subset of S meeting x and M(S), we obtain that S and M(S) are connected. Observing that T is a compact uniquely divisible subsemigroup of S, it will be sufficient for the purpose of proving that T = S to show that $\phi(1) \in T$, since, in this case, $\phi(s) = \phi(1)^s \in T$ for each $s \in Q^+$. Now let $A = M(1)\backslash(M(1) \cap N(1))$ and let $B = N(1)\backslash(M(1) \cap N(1))$. Then $S\backslash(M(1) \cup N(1)) = A \cup B$ and A and B are separated in S. Since T is connected and 1, $e \in T$, with $1 \in N(1)$ and $e \in M(1)$, we have that $T \cap (M(1) \cap N(1)) \neq \square$. Let $y \in T \cap (M(1) \cap N(1))$. Then, since $y \in M(1)$, there exists a net $\{r_\alpha\}$ in Q^+ with $r_\alpha > 1$ for each α such that $\{\phi(r_\alpha)\} \to y$. Now the net $\{\phi(r_\alpha - 1)\}$ clusters to some $z \in S$, so that $\{\phi(r_\alpha)\} = \{\phi(r_\alpha - 1)\phi(1)\}$ clusters to $z\phi(1)$, and we obtain that $y = z\phi(1)$. On the other hand, since $y \in N(1)$, there exists a net $\{s_\beta\}$ in Q^+ such that $s_\beta < 1$ for each β and $\{\phi(s_\beta)\} \to y$. Now the net $\{\phi(1 - s_\beta)\}$ clusters to some $w \in S$, so that $\{\phi(1)\} = \{\phi(1 - s_\beta)\phi(s_\beta)\}$ clusters to xy, and we have $\phi(1) = wy$. Combining this with $y = z\phi(1)$, we obtain that $\phi(1) = wz\phi(1)$. Let $a = wz$, then $\phi(1) = a\phi(1)$ and by recursion $\phi(1) = a^n\phi(1)$. Now if a were not in H(1), we would have that $\{a^n\}$ clusters to e (using (c) so that $\phi(1) = e\phi(1) \in M(S)$, which is easily shown to be a contradiction to the fact that S is not a group, since $M(S) = H(e)$ is uniquely divisible). Thus $wz = a \in H(1)$, and hence, by virtue of 1.34, we have that $w \in H(1)$. We conclude that $\phi(1) = wy \in H(1)T \subset T$, $\phi(1) \in T$, $S = T$, and the proof of (d) is complete.

The proof of (e) is contained in the proof of (d).

To establish (f), we first show that M(S) is solenoidal. Define $\psi : Q^+ \to M(S)$ by $\psi(r) = \phi(r)e$. Then ψ is a dense homomorphism, so that M(S) is unithetic, and it follows from 6.20 that M(S) is solenoidal. Finally to show that H(1) is solenoidal, again notice that both H(1) and $\overline{\sigma(\mathbb{H})}$ are uniquely divisible subsemigroups of S. From (d), we have that $\phi(1) = ht$ for some $h \in H(1)$ and $t \in \overline{\sigma(\mathbb{H})}$. Define

$\psi : Q^+ \to H(1)$ by $\psi(r) = h^r$ for some $r \in Q^+$. Then clearly ψ is a homomorphism. To show that ψ is dense let $h_1 \in H(1)$ and let $\{r_\alpha\}$ be a net in Q^+ so that $\{\phi(r_\alpha)\} \to h_1$. Now the net $\{\psi(r_\alpha)\} = \{h^{r_\alpha}\}$ clusters to some $h_2 \in H(1)$, and by considering subnets we can assume that $\{\psi(r_\alpha)\} \to h_2$. Also the net $\{t^{r_\alpha}\}$ clusters to some $t' \in \sigma(\mathbb{H})$ and again we assume that $\{t^{r_\alpha}\} \to t'$. Now for each α, we have $\phi(r_\alpha) = \phi(1)^{r_\alpha} = (ht)^{r_\alpha} = h^{r_\alpha}t^{r_\alpha} = \psi(r_\alpha)t^{r_\alpha}$, and $\{\phi(r_\alpha)\} \to h_1$, $\{\psi(r_\alpha)\} \to h_2$, and $\{t^{r_\alpha}\} \to t'$, so that $h_1 = h_2t'$. We obtain that $t' = h_2^{-1}h_1 \in H(1) \cap \overline{\sigma(\mathbb{H})}$, so that $t' = 1$ and $h_1 = h_2$. It follows that $\{\psi(r_\alpha)\} \to h_1$ and $H(1)$ is unithetic. Again from 6.20, we have that $H(1)$ is solenoidal. ■

Combining 6.19 and 6.21, we have the following:

6.22 Corollary. Let S be a compact subunithetic semigroup which is not a group. Then S is a connected monoid, $H(1)$ and $M(S)$ are solenoidal groups, $E(S) = \{1,e\}$, where e is the identity of $M(S)$, and there exists a one-parameter semigroup $\sigma : \mathbb{H} \to S$ such that $S = \overline{\sigma(\mathbb{H})}H(1)$.

As a consequence of 6.22, we see that each compact subunithetic semigroup $S = \overline{\sigma(\mathbb{H})}H(1)$ is a cylindrical semigroup since the map $\delta : \sigma(\mathbb{H}) \times H(1) \to S$ defined by $\delta(a,b) = ab$ is clearly a continuous surmorphism. From the fact that $H(1)$ is solenoidal, we have that S is a continuous surmorphic image of $\Phi \times \Sigma$, where $\Phi[\Sigma]$ is the universal compact solenoidal semigroup [group]. It is shown in [Hildebrant, 1969] that $\Phi \times \Sigma$ is the object of the Bohr compactification of Q^+, and hence $\Phi \times \Sigma$ is the universal compact subunithetic semigroup. Rather than proceeding with this development, we turn now to showing that compact subunithetic semigroups constitute the fundamental building blocks of compact divisible semigroups.

6.23 Theorem. Let S be a compact divisible semigroup and let $x \in S$. Then there exists a minimal compact divisible subsemigroup T of S containing x, and T is subunithetic with x as a generator.

Proof. Let $\mathcal{B}$ denote the collection of compact divisible subsemigroups of S containing x. Then $\mathcal{B}$ is partially ordered by inclusion, so that, by virtue of the Hausdorff Maximality Principle, there exists a maximal chain $\mathcal{B}'$ contained in $\mathcal{B}$. Let $T = \cap \mathcal{B}'$. Then T is a compact subsemigroup of S containing x. To see that T is divisible, let $y \in T$ and let $n \in \mathbb{N}$. Then for each $B \in \mathcal{B}'$, there exists $z_B \in B$ such that $z^n = y$. Let z be a cluster point of the net $\{z_B\}$. Then clearly $z^n = y$ and $z \in T$. Thus T is a compact divisible subsemigroup of S containing x and is minimal with respect to this property due to the maximality of $\mathcal{B}'$.

We now proceed to demonstrate that T is subunithetic. Define a sequence $\{x_n\}$ in T recursively so that $x_1 = x$ and $x_n^n = x_{n-1}$ for $n \geqslant 2$ in $\mathbb{N}$. Let $D = \{x_n^m : m, n \in \mathbb{N}\}$. We will show that D is a divisible abelian subsemigroup of T. It is transparent that $D \supset T$ and that $x \in D$. For n, m, p, q $\in \mathbb{N}$, we have that

$$x_p^n x_q^m = x_{pq}^{\left[\frac{n(pq)!}{p!} + \frac{m(pq)!}{q!}\right]}$$

and consequently D is an abelian subsemigroup of T. To see that D is divisible, let $x_n^m \in D$ and let $q \in \mathbb{N}$. Let $p \in \mathbb{N}$ such that $p!/n!q$ is in $\mathbb{N}$. Then

$$\left(x_p^{\frac{mp!}{n!q}}\right)^q = x_p^{\frac{mp!}{n!}} = \left(x_p^{\frac{p!}{n!}}\right)^m = x_n^m$$

and hence D is divisible. Since T is minimal and $\overline{D}$ is divisible, we have that $T = \overline{D}$, and consequently T is abelian. Now let $\pi : T_* \to T$ be the uniquely divisible reflection of T and choose $z \in T_*$ such that $\pi(z) = x$. Define $\phi : Q^+ \to T$ by $\phi(r) = \pi(z^r)$. Then ϕ is a homomorphism with $\phi(1) = x$, and again the minimality of T yields that $T = \phi(Q^+)$. We conclude that T is subunithetic with generator x. ∎

Notice that 6.23 together with 6.21 (e) provides an alternative proof of the fact that a compact semigroup S is divisible if and only if each component of S is a divisible subsemigroup. This same combination gives the following:

6.24 Corollary. Let S be a compact divisible semigroup such that $E(S) = \{0,1\}$, where 0 is a zero for S and 1 is an identity, and such that $S\backslash E(S) \neq \square$. Then S is connected.

If one considers a compact divisible semigroup S in which the subgroups of S are degenerate, e.g., a compact divisible holoid, then, in view of 6.21, we see that the non-degenerate subunithetic semigroups are copies of either I_u (usual) or I_n (nil), and in the uniquely divisible case, only copies of I_u can appear, since I_n is not uniquely divisible. In particular, a compact connected uniquely divisible holoid in which the idempotent are totally disconnected is a union of copies of I_u. We will pursue the study of such semigroups in the next section.

FINITE DIMENSIONAL SEMIGROUPS

Here we discuss a result appearing in [Brown and Friedberg, 1968] and provide an alternate proof as suggested in [Lawson and Madison, 1971].

We will be considering compact connected uniquely divisible semigroups whose idempotent set consists of a zero 0 and an identity 1. Notice that in view of 6.21 and 6.23 the connectedness of such a semigroup S is equivalent to the statement $S\backslash E(S) \neq \square$. One of the distinctive algebraic properties of these semigroups is that they have no proper zero divisors, i.e., $S\backslash\{0\}$ is a subsemigroup of S. In the case that S is a holoid, this fact can be derived from considerations of results in the second section of this chapter. However, as a matter of interest, we provide a proof in a more general setting.

6.25 Theorem. Let S be a compact connected uniquely divisible semigroup with $E(S) = \{0,1\}$. Then $S\backslash\{0\}$ is a subsemigroup of S.

Proof. Suppose that $x, y \in S$, $x \neq 0$, and $xy = 0$. We will demonstrate that $y = 0$. Considering the element

$x^{\frac{1}{2}}yx^{\frac{1}{2}}$, we have $(x^{\frac{1}{2}}yx^{\frac{1}{2}})^2 = (x^{\frac{1}{2}}yx^{\frac{1}{2}})(x^{\frac{1}{2}}yx^{\frac{1}{2}}) = (x^{\frac{1}{2}}y)(xy)x^{\frac{1}{2}} = (x^{\frac{1}{2}}y)0x^{\frac{1}{2}} = 0 = 0^2$, and hence $x^{\frac{1}{2}}yx^{\frac{1}{2}} = 0$ by virtue of unique divisibility. Considering the element $x^{\frac{1}{2^2}}yx^{\frac{1}{2^2}}$, we have $(x^{\frac{1}{2^2}}yx^{\frac{1}{2^2}})^{2^2} = (x^{\frac{1}{2^2}}yx^{\frac{1}{2^2}})(x^{\frac{1}{2^2}}yx^{\frac{1}{2^2}})(x^{\frac{1}{2^2}}yx^{\frac{1}{2^2}})(x^{\frac{1}{2^2}}yx^{\frac{1}{2^2}}) = (x^{\frac{1}{2^2}}y)(x^{\frac{1}{2}}yx^{\frac{1}{2}})(yx^{\frac{1}{2^2}})(x^{\frac{1}{2^2}}yx^{\frac{1}{2^2}}) = (x^{\frac{1}{2^2}}y)0(yx^{\frac{1}{2^2}})(x^{\frac{1}{2^2}}yx^{\frac{1}{2^2}}) = 0 = 0^{2^2}$, and hence $x^{\frac{1}{2^2}} = 0$ by virtue of unique divisibility. Continuing inductively, we obtain that $x^{\frac{1}{2^n}}yx^{\frac{1}{2^2}} = 0$ for each $n \in \mathbb{N}$. In view of the fact that $x \neq 0$, the sequence $\{x^{\frac{1}{2^n}}\}$ clusters to some $h \in H(1)$, and we have $hyh = 0$. It follows that $y = 1y1 = h^{-1}hyhh^{-1} = h^{-1}0h^{-1} = 0$. ∎

For the proof of the main result of this section, we employ the following lemma:

6.26 Lemma. Let S be a compact connected uniquely divisible holoid with $E(S) = \{0,1\}$. Then $\exp|C(1) : C(1) \to S\backslash\{0\}$ is a topological isomorphism.

Proof. In view of 6.3 (d) and 6.7, we see that $\exp : \mathrm{Hom}(\mathbb{H}^0,S) \to S$ is a topological isomorphism. Combining 6.23 and 6.21 (d), we see that $\exp C(1) = S\backslash\{0\}$ and that $\mathrm{Hom}(\mathbb{H}^0,S)\backslash C(1) = \{f\}$, where $f(x) = 0$ for each $x \in \mathbb{H}^0$. The conclusion follows. ∎

For $n \in \mathbb{N}$, let E^n denote Euclidean n-space as a topological vector space over $\mathbb{R}$.

The next result is found in [Brown and Friedberg, 1968].

6.27 Theorem. Let S be a compact connected uniquely divisible n-dimensional holoid with $E(S) = \{0,1\}$. Then S is topologically isomorphic to the one point compactification of a closed proper cone in E^n, in which the ideal point acts as a zero.

Proof. In view of 6.26, we see that $C = S\backslash\{0\}$ is a locally compact ordered topological cone semigroup, so that, by virtue of 6.18, there is an embedding of C into a topological vector space V over $\mathbb{R}$. Without loss of generality, we can assume that $C \subset V$ and that $V = C - C$. Let B be a maximal linearly independent subset of C. Then B generates C, which in turn generates V, and hence B is a basis for V.

Suppose now, for the purpose of contradiction, that B contains n + 1 independent elements of V say $v_1,\ldots,v_{n+1}$. Let $I = [0,1]$ and let I^{n+1} denote the n+1-fold Cartesian product of I with itself as a space. Define $\psi : I^{n+1} \to C$ by $\psi(\alpha_1,\alpha_2,\ldots,\alpha_{n+1}) = \sum_{j=1}^{n+1} \alpha_j v_j$. Then ψ is a homeomorphism of I^{n+1} onto a subspace of C, contradicting the fact that C is n-dimensional. Thus V is linearly topologically isomorphic to E^m for some m with $m \leqslant n$ and C is a closed proper cone in E^m, so that in fact $m = n$. The conclusion now follows. ∎

FURTHER RESULTS ON DIVISIBLE SEMIGROUPS

Here we present an expository discussion of further results pertaining to compact divisible semigroups and in some cases their algebraic predecessors. We begin with a result which extends the class of semigroups considered in 6.27. It appears in [Brown and Friedberg, 1968].

> *6.28 Theorem.* Let S be a compact connected finite dimensional uniquely divisible abelian semigroup such that $E(S) = \{0,1\}$. Then S is topologically isomorphic to $((S/H(1)) \times H(1))/(\{0\} \times H(1))$.

There are additional results pertaining to representations of compact uniquely divisible semigroups appearing both in [Brown and Friedberg, 1968] and [Brown and Friedberg, 1969]. One such result is the following:

> *6.29 Theorem.* Let S be a compact finite dimensional uniquely divisible abelian semigroup such that E(S) is totally disconnected. Then Hom $(S,\mathbb{D})$ separates points.

Combining 6.14, 6.15, and 6.18 with some results of the second section we obtain the following generalization of 6.27. It appears in [Keimel, 1971].

6.30 Theorem. Let S be a compact uniquely divisible holoid with $E(S) = \{0,1\}$. Then there is a proper locally compact cone semigroup C in a real topological vector space such that S is topologically isomorphic to $C \cup \{\infty\}$ (the one-point compactification of C with the ideal point ∞ acting as a zero).

An algebraic predecessor of 6.30 and of 6.27 appears in [Brown and LaTorre, 1966].

One of the more significant early contributions of the theory of compact divisible semigroups is the so-called "backward gamma technique" introduced in [Hudson, 1959] and further explored in [Hofmann, 1960]. This led to the results of the section on subunithetic semigroups and eventually to 6.27 and relating results. This technique was employed in [Hildebrant, 1967] to characterize those uniquely divisible abelian semigroups on the two-cell whose idempotent set consists of a zero and an identity. Nonabelian semigroups satisfying these conditions were characterized through a sequence of papers including [DeVun, 1970], [Borrego, Cohen, and DeVun, 1971], and [DeVun, 1972].

It has been conjectured in [Brown and Friedberg, 1969] that each compact (abelian) semigroup can be embedded in a compact (abelian) divisible semigroup. This conjecture, along with the facts that many compact semigroups contain an abundance of compact divisible subsemigroups and that irreducible semigroups are divisible, has provided substantial motivation for the study of compact divisible semigroups. A partial solution of the conjecture is given in the following. It appears in [Hildebrant and Lawson, 1972].

6.31 Theorem. Let S be a compact power-cancellative abelian semigroup such that $\{x^n : x \in S\}$ is an ideal of S for each $n \in \mathbb{N}$. Then S can be embedded in a compact uniquely divisible abelian semigroup.

The main technique used to establish 6.31 was a topological version of the technique used to prove the algebraic predecessor appearing in [Hancock, 1960]. Employment of this technique was made possible by the quotient semigroup result in [Lawson and Madison, 1971] (see 1.56).

An additional partial solution of the compact divisible embedding problem appears in [Hildebrant, 1976].

6.32 Theorem. Each compact totally disconnected abelian semigroup can be embedded in a compact divisible abelian semigroup.

SYMBOLS

BIBLIOGRAPHY

Aczel, J., and Wallace, A. D., "A note on generalizations of transitive systems of transformations" Colloq. Math., 17 (1967), 29-34.

Anderson, James A., "Characters of commutative semigroups. II. Topological Semigroups", Math. Sem. Notes Kobe Univ., 7 (1979), 491-498.

Anderson, L. W., and Hunter, R. P., "The H-equivalence in compact semigroups" Bull. de la Soc. Math. de Belg., 14 (1962a), 274-296; "Homomorphisms and dimension" Math. Ann., 147 (1962b), 248-268; "The H-equivalence in compact semigroups II" J. Austr. Math. Soc., 3 (1963a), 288-293; "Small continua at certain orbits" Archiv. der Math., 14 (1963b), 350-353; "Sur les espaces fibres associes a une D-classe d'un demi-groupe compact" Bull. Acad. Polon. Sci. Ser. Math. Astronom. Phys., 12 (1964a), 249-251; "Sur les demi-groupes compacts et connexes" Fund. Math., 56 (1964b), 183-187; "Certain homomorphisms of a compact semigroup into a thread" J. Austral. Math. Soc., 7 (1967), 311-322; "On the compactification of certain semigroups" Proc. Intn. Symp. on Extension Theory of Topological Structures, VEB, Berlin, 1969; "Compact semigroups having certain one dimensional hyperspaces" Amer. J. Math., 92 (1970), 894-896; "A remark on compact semigroups having certain decomposition spaces embeddable in the plane" Bull. Austral. Math. Soc., 4 (1971a), 137-139; "On the continuity of certain homomorphisms of compact semigroups" Duke Math. J., 38 (1971b), 409-414; "A remark on finite dimensional compactifications" Duke Math. J., 38 (1971c),

605-607; "Compact semigroups having certain hyperspaces in the plane" Bull. Aus. Math. Soc., 4 (1971d), 137-139; "On the infinite subsemigroups of certain compact semigroups" Fund. Math., 74 (1972), 1-19; "A remark on finite dimensional compact connected monoids" Proc. Amer. Math. Soc., 42 (1973a), 602-606; "Homomorphisms having a given H-class as a single class" J. Aus. Math. Soc., 15 (1973b), 7-14; "A remark on finite dimensional compact connected monoids" Proc. Amer. Math. Soc., 42 (1974), 602-606.

Anderson, L. W., Hunter, R. P., and Koch, R. J., "Some results on stability in semigroups" Trans. Amer. Math. Soc., 117 (1965), 521-529.

Anderson, L. W., and Ward, L. E., Jr., "One-dimensional topological semilattices" Ill. J. Math., 5 (1961), 182-186.

Baker, J. W., and Butcher, R. J., "The Stone Cech compactification of a topological semigroup", Math. Proc. Cambridge Philos. Soc., 80 (1976), 103-107.

Balman, R., Continua Acting on the Unit Interval, Dissertation, Florida Univ., 1966.

Bednarek, A. R., and Wallace, A. D., "Relative ideals and their complements I" Rev. Roum. Math. Pure et Appl., 10 (1966), 15-22; "Finite approximants of compact totally disconnected acts" Math. Systems Theory, 1 (1967a), 209-216; "A relation-theoretic result with applications in topological algebra" Math. Systems Theory 1 (1967b), 217-224; "Equivalences on machine state spaces" Mat. isasopis, 17 (1967c), 1-7.

Berglund, J. F., "Compact semitopological inverse Clifford semigroups" Semigroup Forum, 5 (1973), 191-215.

Berglund, J., and Hofmann, K. H., Compact Semitopological Semigroups and Weakly Almost Periodic Functions, Springer-Verlag series, 1967.

Berglund, J. F., Junghenn, H. D., and Milnes, P., "Universal mapping properties of semigroup compactifications" Semigroup Forum, 15 (1978a), 375-386; Compact Right Topological Semigroups and

Generalizations of Almost Periodicity. Lecture Notes in Mathematics; Springer-Verlag, New York, 1978b.

Bertman, M. O., "Free topological inverse semigroups" Semigroup Forum, 8 (1974), 266-269.

Bertman, M. O., and West, T. T., "Conditionally compact bicyclic semitopological semigroups" Proc. Roy. Irish Acad. Sec., 76 (1976), 219-226.

Borrego, J. T., "Homomorphic retractions in semigroups" Proc. Amer. Math. Soc., 18 (1967), 716-719; "Adjunction semigroups" Bull. Austral. Math. Soc., 1 (1969a), 47-58; "Point-transitive actions by a standard metric thread" Proc. Amer. Math. Soc., 23 (1969b), 261-265; "Point-transitive actions by certain abelian real clans" Aequ. Math., 5 (1970a), 255-259; "Continuity of operations in a semilattice" Colloq. Math., 21 (1970b), 49-52.

Borrego, J. T., Cohen, H., and DeVun, E., "Uniquely representable semigroups II" Pac. J. Math., 39 (1971), 573-579.

Borrego, J. T., and DeVun, E., "Point-transitive actions by the unit interval" Can. J. Math., 22 (1970), 255-259; "Selections and unitary actions of semigroups" Kyungpook Math. J., 11 (1971), 9-12; "Maximal semigroup orbits" Semigroup Forum, 4 (1972), 61-68.

Bourbaki, N., Topologie generale. Actual. Sci. et Indust., Hermann, Paris, 1951.

Bowman, T. T., "A construction principle and compact Clifford semigroups" Semigroup Forum, 2 (1971), 343-353; "Generators of categories of compact irreducible semigroups" Semigroup Forum, 5 (1973), 331-339; "Generators for categories of compact irreducible semigroups" Semigroups Forum, 5 (1972/73), 331-339.

Brown, D. R., "On clans of non-negative matrices" Proc. Amer. Math. Soc., 15 (1964), 671-674; "Topological semilattices on the two-cell" Pac. J. Math., 15 (1965), 35-46; "Matrix representations of compact simple semigroups" Duke Math. J., 33 (1966), 69-74.

Brown, D. R., and Friedberg, M., "Representation theorems for uniquely divisible semigroups" Duke Math. J., 35 (1968), 341-352; "A new notion of semicharacters" Trans. Amer. Math. Soc., 141 (1969), 387-401; "A survey of compact divisible commutative semigroups" Semigroup Forum, 1 (1970a), 143-161; "Representations of topological semigroups" Met. Casopis. Solven. Akad. Vied., 20 (1970b), 304-314; "Linear representations of certain compact semigroups" Trans. Amer. Math. Soc., 160 (1971), 453-465.

Brown, D. R., and LaTorre, J. G., "A characterization of uniquely divisible commutative semigroups" Pac. J. Math., 18 (1966), 57-60.

Brown, D. R., and Stepp, J. W., "Inner points and breadth in certain compact semilattices" Proc. Amer. Math. Soc. (to appear).

Brown, D. R., and Stralka, A. R., "Problems about compact semilattices" Semigroup Forum, 6 (1973), 265-270; "Compact totally instable zero-dimensional semilattices" General Topology and Appl., 7 (1977), 151-159.

Burckel, R. B., Weakly Almost Periodic Functions on Semigroups. Gordon and Breech, New York, 1971.

Carruth, J. H., and Clark, C. E., "Representations of certain compact semigroups by HL-semigroups" Trans. Amer. Math. Soc., 149 (1970), 327-337; "Compact totally H-ordered semigroups" Proc. Amer. Math. Soc., 27 (1971), 199-204; "Concerning restrictions of Green's H-equivalence" Semigroup Forum, 5 (1972), 186-189; "H-coextensions in the category of compact semigroups" Semigroup Forum, 7 (1974), 164-179; "Generalized Green's theories" Semigroup Forum, 20 (1979), 95-127.

Carruth, J. H., Hofmann, K. H., and Mislov, M. W., "Errors in elements of compact semigroups" Semigroup Forum, 5 (1973), 285-322.

Carruth, J. H., and Lawson, J. D., "Semigroups through semilattices" Trans. Amer. Math. Soc., 152 (1970a), 597-608; "On the existence of one-parameter semigroups" Semigroup Forum, 1 (1970b), 85-90.

Cartan, E., La Topologie des Groupes de Lie. Hermann, Paris, 1936.

Chevalley, C., Theory of Lie Groups. Princeton Univ. Press, Princeton, N. J., 1946.

Christoph, F. T., Jr., "Ideal extensions of topological semigroups" Can. J. Math., 22 (1970a), 1168-1175; "Free topological semigroups in topological groups" Pac. J. Math., 34 (1970b), 343-353; "Embedding topological semigroups in topological groups" Semigroup Forum, 1 (1970c), 224-231; "Factor semigroups which are topological semigroups" Math. Japon., 17 (1972), 21-31.

Clark, C. E., "Locally algebraically independent collections of subsemigroups of a semigroup" Duke Math. J., 35 (1968), 843-851; "Monotone homomorphisms of compact semigroups" J. Austral. Math. Soc., 9 (1969), 167-175; "Certain types of congruences on compact commutative semigroups" Duke Math. J., 34 (1970), 95-101.

Clifford, A. H., "Extensions of semigroups" Trans. Amer. Math. Soc., 68 (1950), 165-173; "Naturally totally ordered commutative semigroups" Amer. J. Math., 76 (1954), 631-646; "Totally ordered commutative semigroups" Bull. Amer. Math. Soc., 64 (1958a), 305-316; "Connected ordered topological semigroups with idempotent endpoints I" Trans. Amer. Math. Soc., 88 (1958b), 80-98; "Connected ordered topological semigroups with idempotent endpoints II" Trans. Amer. Math. Soc., 91 (1959), 193-208.

Clifford, A. H., and Preston, G. B., The Algebraic Theory of Semigroups I. Math. Surveys, 7, Amer. Math. Soc., 1961; The Algebraic Theory of Semigroups II. Math. Surveys, 7, Amer. Math. Soc., 1967.

Cohen, H., "A cohomological definition of dimension for locally compact Hausdorff spaces" Duke Math. J., 21 (1954), 209-224; "A clan with zero without the fixed point property" Proc. Amer. Math. Soc., 11 (1960), 937-939; "When is the kernel closed?" Niew. Archief. voor Wiskunde, 21 (1973), 164-167; "Compact connected semilattices without the fixed point property" Semigroup Forum, 7 (1974), 375-379.

Cohen, H., and Collins, H. S., "Affine semigroups" Trans. Amer. Math. Soc., 93 (1959), 97-113.

Cohen, H., and Koch, R. J., "Acyclic semigroups and multiplications on two-manifolds" Trans. Amer. Math. Soc., 118 (1965), 420-427; "Idempotent semigroups generated by threads I" Semigroup Forum, (to appear).
23 (1981), 247-254. "Idempotents semigroups generated by threads II", Semigroup Forum, 24 (1982), 373-383.

Cohen, H., and Krule, I. S., "Continuous homomorphic images of real clans with zero" Proc. Amer. Math. Soc., 10 (1959), 106-108.

Cohen, H., and Wade, L. I., "Clans with zero on an interval" Trans. Amer. Math. Soc., 88 (1958), 523-535.

Collins, H. S., "Idempotents in the semigroup of measures" Duke Math. J., 27 (1960), 397-400; "The kernel of a semigroup of measures" Duke Math. J., 28 (1961), 387-392; "Remarks on affine semigroups" Pac. J. Math., 12 (1962), 449-455; "Characterizations of convolution semigroups of measures" Pac. J. Math., 14 (1964), 479-492.

Collins, H. S., and Koch, R. J., "Regular D-classes in measure semigroups" Trans. Amer. Math. Soc., 105 (1962), 21-31.

Day, J. M., Algebraic Theory of Machine Languages. Expository Lectures on Topological Semigroups. Academic Press, 1968; "Semigroups acts, algebraic and topological" Proceedings of the Second Florida Symposium on Automata and Semigroups. University of Florida, Gainsville, 1971.

Day, J. M., and Hofmann, K. H., "Clans acts and codimension" Semigroup Forum, 4 (1972), 206-214.

Day, J. M., and Wallace, A. D., "Semigroups acting on continua" J. Anst. Math. Soc., 7 (1967a), 327-340; "Multiplication induced in the state-space of an act" Math. Systems Theory, 1 (1967b), 305-314.

Deleeuw, K., and Glicksberg, I., "Almost periodic functions on semigroups" Acta. Math., 105 (1961), 99-140.

DeVun, E., "Special semigroups on the two-cell" Pac. J. Math., 34 (1970), 639-645; "The equivalence of uniquely divisible semigroups and uniquely representable semigroups on the two-cell" Semigroup Forum, 4 (1972a), 69-72; "Semigroups on the disk with thread boundaries" Kyungpook Math. J., 12 (1972b), 159-169; "Product semigroups" J. Austral. Math. Soc. Ser. A, 22 (1976), 391-399.

Dugundji, J., Topology. Allyn and Bacon, Inc., Boston, 1968.

Eberhart, C., and Selden, J., "On the closure of the bicyclic semigroup" Trans. Amer. Math. Soc., 144 (1969), 115-126.; "One-parameter inverse semigroups" Trans. Amer. Math. Soc., 168 (1972), 53-66.

Eckmann, B., "Uber monothetische Gruppen" Com. Math. Helv., 16 (1943-44), 249-263.

Eilenberg, S., "Sur les transformations continues d'espaces metriques compacts" Fund. Math., 22 (1934), 292-296.

Eilenberg, S., and Steenrod, N., Foundations of Algebraic Topology. Princeton Univ. Press, Princeton, N. J., 1952.

Ellis, R., "Locally compact transformation groups" Duke Math. J., 24 (1957a), 119-125; "A note on the continuity of the inverse" Proc. Amer. Math. Soc., 8 (1957b), 372-373.

Farley, R. W., "Positive Clifford semigroups on the plane" Trans. Amer. Math. Soc., 151 (1970), 353-369; "Construction of positive commutative semigroups on the plane", Bull. Inst. Math. Acad. Science 7 (1979), 357-362.

Faucett, W. M., "Compact semigroups irreducibly connected between two idempotents" Proc. Amer. Math. Soc., 6 (1955a), 741-747; "Topological semigroups and continua with cut points" Proc. Amer. Math. Soc., 6 (1955b), 748-756.

Fort, M. K., "Homogeneity of infinite products of manifolds with boundary" Pac. J. Math., 12 (1962), 879-884.

Friedberg, M., "On representations of certain semigroups" Pac. J. Math., 19 (1966), 269-274; "Metrizable approximations of semigroups" Colloq. Math., 25 (1972a), 63-69; "Some examples of clans" Semigroup Forum, 4 (1972b), 156-164; "On compactifying semigroups" Semigroup Forum, 10 (1975), 39-54; "Compactifications of finite-dimensional cones" Semigroup Forum, 15 (1978a), 199-228; "On the universal compactification of a cone" Rocky Mtn. J., 8 (1978b), 503-526; "Almost-periodic functions, compactifications, and faces of finite-dimensional cones" Math. Zeit., 176 (1981), 53-61.

Friedberg, M., and Stepp, J. W., "A note on the Bohr compactification" Semigroup Forum, 6 (1973), 362-364.

Fulp, R. O., and Stepp, J. W., "Semigroup extensions" J. Reine Angew. Math., 248 (1971a), 28-41; "The semigroup of semigroup extensions" Semigroup Forum, 2 (1971b), 173-180; "Structure of the semigroup of semigroup extensions" Trans. Amer. Math. Soc., 158 (1971c), 63-73.

Gelbaum, B., Kalisch, G. K., and Olmstead, J. M. H., "On the embedding of topological semigroups and integral domains" Proc. Amer. Math. Soc., 2 (1951), 807-821.

Gierz, G., Hoffmann, K. H., Keimel, K., Lawson, J. D., Mislove, M., and Scott, D., A Compendium of Continuous Lattices. Heilelberg, 1980.

Gleason, A. M., "Arcs in locally compact groups" Proc. Nat. Acad. Sci., 36 (1950a), 663-667; "Spaces with a compact Lie group of transformations" Proc. Amer. Math. Soc., 1 (1950b), 35-43; "Groups without small subgroups" Ann. of Math., 56 (1952), 193-212.

Glicksberg, I., "Convolution semigroups of measures" Pac. J. Math., 9 (1959), 51-67.

Graham, G., Differentiability and Semigroups. Dissertation, University of Houston, 1979.

Halmos, P. R., and Samelson, H., "On monothetic groups" Proc. Nat. Acad. Sci. U.S.A., 28 (1942), 254-258.

Hancock, V. R., Commutative Schreier Extensions of Semigroups. Doctoral Dissertation, Tulane University, 1960.

Hanson, T. H. McH., "Actions of a locally compact group with zero"

Can. J. Math., 23 (1971a), 413-420; "Products of locally compact groups with zero and their actions" Semigroup Forum, 3 (1971b), 180-184; "Actions of a locally compact group with zero II" Semigroup Forum, 3 (1972a), 371-374; "Actions that fiber and vector semigroups" Can. J. Math., 24 (1972b), 29-37.

Hanson, T. H. McH., and King, L., "Noncompact semigroup actions" Semigroup Forum, 7 (1974), 58-73.

Hays, T. E., "Irreducible semigroups are monotone" Semigroup Forum, 10 (1975), 25-31; "Trees and monotone structures" Semigroup Forum, 13 (1977), 377-383.

Hewitt, E., "Compact monothetic semigroups" Duke Math. J., 23 (1956), 447-457.

Hewitt, E., and Ross, K. A., Abstract Harmonic Analysis I. Academic Press, New York, 1963.

Hildebrant, J. A., "On compact unithetic semigroups" Pac. J. Math., 21 (1967a), 265-273; "On uniquely divisible semigroups on the two-cell" Pac. J. Math., 23 (1967b), 91-95; "On compact

Gottinger, H. W., "Existence of a utility on a topological semigroup", Theory and Decision, 7 (1976), 148-158.

divisible abelian semigroups" Proc. Amer. Math. Soc., 19 (1968), 405-410; "The universal compact subunithetic semigroup" Proc. Amer. Math. Soc., 23 (1969), 220-224; "A note on the uniquely divisible reflection of a compact abelian semigroup" An. Acad. brasil Cienc., 43 (1971), 37-38; "Compact p-adic semigroups" Duke Math. J., 39 (1972), 39-44; "Extending congruences on factors of ideally compactifiable semigroups" Semigroup Forum, 12 (1976), 245-250; "Projective topological semigroups", An. Acad. brasil Cienc., 53 (1981), 677-681.

Hildebrant, J. A., and Lawson, J. D., "Embedding in compact uniquely divisible semigroups" Semigroup Forum, 4 (1972), 295-300; "The Bohr compactification of a dense ideal in a topological semigroup" Semigroup Forum, 6 (1973), 86-92.

Hildebrant, J. A., Lawson, J. D., and Yeager, D. P., "The translational hull of a topological semigroup" Trans. Amer. Math. Soc., 221 (1976), 251-280.

Hinman, B., "Products of threads" Semigroup Forum, 8 (1974), 1-20.

Hireschorn, R., "Topological semigroups, sets of generators, and controllability" Duke Math. J., 40 (1973), 937-947.

Hocking, J. G., and Young, G. S., Topology. Addison-Wesley, Reading, 1961.

Hofmann, K. H., "Topologische Halbgruppen mit dichter submonoger Unterhalbgruppe" Math. Zeit., 74 (1960), 232-276; "Locally compact semigroups in which a subgroup with compact complement is dense" Trans. Amer. Math. Soc., 106 (1963a), 19-51; "Homogeneous locally compact groups with compact boundary" Trans. Amer. Math. Soc., 106 (1963b), 52-63; "Topological semigroups; History, Theory, Applications" Uber. Deutch. Math. Verein, 78 (1976), 9-59.

Hofmann, K. H., and Hunter, R. P., "On a centralizer result of Hunter" Semigroup Forum, 6 (1973), 365-372.

Hofmann, K. H., and Lawson, J. D., "The local theory of semigroups in nilpotent Lie groups" Semigroup Forum, 23 No. 4 (1981), 343-357; "Foundations of Lie Semigroups" (to appear).

Hofmann, K. H., and Mislove, M., "The centralizing theorem for left normal groups of units in compact monoids" Semigroup Forum, 3 (1971), 31-42.

Hofmann, K. H., Mislove, M. W., and Stralka, A. R., "Dimension raising maps in topological algebra" Math. Zeit., 135 (1973), 1-36.

Hofmann, K. H., and Mostert, P. S., "Irreducible semigroups" Bull. Amer. Math. Soc., 70 (1964a), 621-627; "Totally ordered D-class decompositions" Bull. Amer. Math. Soc., 70 (1964b), 765-771; "Connected extensions of simple semigroups" Czeh. Math. J., 15 (1965c), 295-298; Elements of Compact Semigroups. Merrill Books, Inc., Columbus, Ohio, 1966; "Problems about compact semigroups" Semigroups, Proc. Wayne State Univ. Symp. 1968, Academic Press, New York, 1969.

Hofmann, K. H., and Stralka, A. R., "Mapping cylinders and compact monoids" Math. Ann., 205 (1973a), 219-239; "Push-outs and strict projective limits of semilattices" Semigroup Forum, 5 (1973b), 243-261.

Hoo, C. S., and Shum, K. P., "On the nilpotent elements of semigroups" Colloq. Math., 25 (1972), 211-224; "Ideals and radicals in topological semigroups", Southeast Asian Bull. Math., 3 (1979), 140-150.

Horne, J. G., Jr., "Real commutative semigroups on the plane" Pac. J. Math., 11 (1961), 981-997; "One-parameter subgroups in semigroups in the plane" Mich. Math. J., 9 (1962a), 177-186; "Semigroups on a half-plane" Trans. Amer. Math. Soc., 105 (1962b), 9-20; "Real commutative semigroups on the plane II" Trans. Amer. Math. Soc., 104 (1962c), 17-23; "The boundary of a one-parameter group in a semigroup" Duke Math. J., 31 (1964), 109-117; "A locally compact connected group acting on the plane has a closed orbit" Ill. J. Math., 9 (1965), 644-650; "Semigroups on a half-space" Trans. Amer. Math. Soc., 147 (1970), 1-53.

Houston, R., Cancellative Semigroups on Manifolds. Dissertation, University of Houston, 1973.

Hudson, A. L. (see also Lester), "On the structure of semigroups on a non-compact manifold" Mich. Math. J., 8 (1961a), 11-19; "Some semigroups on a n-cell" Trans. Amer. Math. Soc., 99 (1961b), 255-263; "Example of a non-acyclic continuum semigroup S with 0 and S = ESE" Proc. Amer. Math. Soc., 14 (1963), 648-653.

Hudson, A. L., and Mostert, P. S., "A finite dimensional homogeneous clan is a group" Annals of Math., 78 (1963), 41-46.

Hunter, R. P., "On the semigroup structure of continua" Trans. Amer. Math. Soc., 93 (1959), 356-368; "Certain upper semi-continuous decompositions of a semigroup" Duke Math. J., 27 (1960), 283-290; "Note on arcs in semigroups" Fund. Math., 49 (1961a), 233-245; "Certain homomorphisms of compact connected semigroups"

Duke Math. J., 28 (1961b), 83-88; "On a conjecture of Koch" Proc. Amer. Math. Soc., 12 (1961c), 138-139; "On one-dimensional semigroups" Math. Ann., 146 (1962), 383-396; "On the structure of homogroups with applications to the theory of compact connected semigroups" Fund. Math., 52 (1963), 62-102; "Compact semigroups with low dimensional orbit spaces" Proc. Amer. Math. Soc., 40 (1973), 277-279.

Hunter, R. P., and Rothman, N. J., "Characters and cross sections for certain semigroups" Duke Math. J., 29 (1962), 347-366.

Hurewicz, W., and Wallman, H., Dimension Theory. Princeton Univ. Press, 5th Ed., Princeton, N. J., 1959.

Iseki, K., "On compact abelian semigroups" Mich. Math. J., 2 (1953), 59-60; "On compact semigroups" Proc. Japan Acad., 32 (1956), 221-224.

Iwassawa, K., "Finite and compact groups" Sugaku, 1 (1948), 30-31; "On some types of topological groups" Ann. of Math., 50 (1949), 507-558.

Jonsdottir, K. H., "Ideal compactifications in uniquely divisible semigroups" Semigroup Forum, 14 (1977), 355-374; "Kernel compactifications of uniquely divisible semigroups" Semigroup Forum, 15 (1978), 269-279.

Junghenn, H. D., "The Mask compactification of a dense subsemigroup", Semigroup Forum, 17 (1979), 261-265.

Keimel, K., "Eine Exponentialfunktion fur kompakte abelsche Halbgruppen" Math. Zeit., 96 (1967a), 7-25; "Lokal kompakte Kegelhalbgruppen und deren Einbettung in topologische Vektorraume" Math. Zeit., 99 (1967b), 405-428; "Congruence relations on cone semigroups" Semigroup Forum, 3 (1971), 130-147.

Keleman, C. F., "Transitive semigroup actions" Trans. Amer. Math. Soc., 146 (1969), 369-375; "Actions by a bisimple w-semigroup" Semigroup Forum, 4 (1972), 248-251; "Recursions with uniquely determined topologies" Fund. Math. 77 (1973), 219-225.

Kelley, J. L., General Copology, D. Van Nostrand Co., Inc., Princeton, N. J., 1955.

King, L., "Actions of a noncompact semigroup with zero" Semigroup Forum, 4 (1972), 224-231.

Kharashani, H., "Left thick subsets of a topological semigroup", Ill. J. Math., 22 (1978), 41-48.

Koch, R. J., On Topological Semigroups. Dissertation, Tulane Univ., 1953; "On monothetic semigroups" Proc. Amer. Math. Soc., 8 (1957a), 397-401; "Note on weak cutpoints in clans" Duke Math. J., 24 (1957b), 611-616; "Arcs in partially ordered spaces" Pac. Math. J., 9 (1959), 723-728; "Ordered semigroups in partially ordered semigroups" Pac. Math. J., 10 (1960), 1333-1336; "Threads in compact semigroups" Math. Zeit., 86 (1964), 312-316; "Compact connected spaces supporting topological semigroups" Semigroup Forum, 1 (1970), 95-102.

Koch, R. J., and Krule, I. S., "Weak cutpoint ordering on hereditarily unicoherent continua" Proc. Amer. Math. Soc., 11 (1960), 679-681.

Koch, R. J., and McAuley, L. F., "Semigroups on trees" Fund. Math., 50 (1962), 341-346; "Semigroups on continua ruled by arcs" Fund. Math. 56 (1964), 1-8.

Koch, R. J., and Wallace, A. D., "Maximal ideals in compact semigroups" Duke Math. J., 21 (1954), 681-686; "Stability in semigroups" Duke Math. J., 24 (1957), 193-195; "Admissibility of semigroup structures on continua" Trans. Amer. Math. Soc., 88 (1958), 277-287; Notes on inverse semigroups" Rev. Roum. Math. Pure et Appl., 9 (1964), 19-24.

Krohn, K., and Rhodes, J., "Algebraic theory of machines I. Prime Decomposition theorem for finite semigroups and machines" Trans. Amer. Math. Soc., 116 (1965), 450-464; "Complexity of finite semigroups" Ann. of Math., 88 (1968), 128-160.

Lau, A., "Concerning costability of compact semigroups" Duke Math. J., 39 (1972a), 644-657; "Small semilattices" Semigroup Forum,

4 (1972b), 150-155; "Coverable semigroups" Proc. Amer. Math. Soc., 28 (1973), 661-664; "Finitely neighborable groupoids" Kyungpook Math. J., 16 (1976), 155-159.

Lawson, J. D., "Topological semilattices with small semilattices" J. London Math. Soc., 1 (1969), 719-724; "Lattices with no interval homomorphisms" Pac. J. Math., 32 (1970a), 459-465; "The relation of breadth and codimension in topological semilattices" Duke Math. J., 37 (1970b), 207-212; "The relation of breadth and codimension in topological semilattices" Duke Math. J., 38 (1971), 555-559; "Dimensionally stable semilattices" Semigroup Forum, 5 (1972), 181-185; "Joint continuity in semitopological semigroups" Ill. Math. J., 18 (1974), 275-285; "Additional notes on continuity in semitopological semigroups" Semigroup Forum, 12 (1976), 265-280; "Compact semilattices which must have a basis of subsemilattices" J. London Math. Soc. (2), 16 (1977), 367-371.

Lawson, J. D., Liukkonen, J. R., and Mislove, M., "Measure algebras of semilattices with finite breadth" Pac. J. Math., 69 (1977), 125-139.

Lawson, J. D., and Madison, B. L., "Peripheral and inner points" Fund. Math., 69 (1970a), 253-266; "Peripherality in semigroups" Semigroup Forum, 1 (1970b), 128-142; "On congruences and cones" Math. Zeit., 120 (1971), 18-24; "Quotients of k-semigroups" Semigroup Forum, 9 (1974), 1-18.

Lawson, J. D., and Williams, W. W., "Topological semilattices and their underlying spaces" Semigroup Forum, 1 (1970), 209-223.

Lester, A. (see also Hudson), "Some semigroups on the two-cell" Proc. Amer. Math. Soc., 10 (1959), 648-655.

L'Heureux, J. E., "The min cone over the circle group" Proc. Amer. Math. Soc., 41 (1973), 625-628.

Lin, Y. F., "On input semigroups of automata" Math Systems Theory, 4 (1970), 35-39.

Los, J., and Schwarz, S., "Remarks on compact semigroups" Colloq. Math., 6 (1958), 265-270.

Luther, I. S., "Uniqueness of the invariant mean on an abelian semigroup" Ill. J. Math., 3 (1959), 28-44; "Uniqueness of the invariant mean on abelian topological semigroups" Trans. Amer. Math. Soc., 104 (1962), 403-411.

Madison, B. L., "Semigroups on coset spaces" Duke Math. J., 36 (1969a), 61-64; "A note on local homogeneity and stability" Fund. Math., 66 (1969b), 123-127; "Quotients of locally compact semigroups" Semigroup Forum, 3 (1981), 277-281.

Madison, B. L., and Seldon, J., "Clans on group-supporting spaces" Proc. Amer. Math. Soc., 18 (1967), 540-545.

Madison, B. L., and Stepp, J. W., "Inversion and joint continuity in semigroups on k-omega spaces" Semigroup Forum, 15 (1978), 195-198.

Marxen, D., "The topology of the free topological semigroup", Coll. Math., 34 (1975-76), 7-16.

McGranery, C., "Some theorems concerning semigroup actions on Euclidean spaces" Semigroup Forum, 8 (1974), 82-88.

McMaster, T. B. M., "Product theorems for generalized topological semilattices" J. London Math. Soc., 13 (1976), 41-52.

Milnes, P., "Compactifications of semitopological semigroups" J. Austr. Math. Soc., 15 (1973a), 488-503; "The weakly almost periodic compactification; another approach" Semigroup Forum, 6 (1973b), 340-345.

Mislove, M. W., "The existence of Irr(X)" Semigroup Forum, 1 (1970), 243-248; "The existence and structure of Irr(X)" Semigroup Forum, 4 (1972), 1-33.

Mitchell, B., Theory of Categories. Academic Press, New York, 1965.

Mitchell, T., "Topological semigroups and fixed points" Ill. J. Math., 14 (1970), 630-641.

Montgomery, D., "Continuity in topological groups" Bull. Amer. Math. Soc., 42 (1936), 879-882.

Montgomery, D., and Zippin, L., Topolotical Transformation Groups, Inters. Tracts 1, Wiley and Son, New York, 1955.

Montgomery, D., and Yang, C. T., "The existence of a slice" Ann. of

Math., 65 (1957), 108-116.

Mostert, P. S., "Local cross sections in locally compact groups" Proc. Amer. Math. Soc., 4 (1953), 645-649; "Plane semigroups" Trans. Amer. Math. Soc., 103 (1962), 320-328; "Untergruppen von Halbgruppen" Math. Zeit., 82 (1963a), 29-36; "A fibering theorem for topological semigroups" Proc. Amer. Math. Soc., 14 (1963b), 87-88; "Continua meeting an orbit at a point" Fund. Math., 52 (1963c), 319-321; "The structure of topological semigroups revisited" Bull. Amer. Math. Soc., 72 (1966), 601-618.

Mostert, P. S., and Shields, A. L., "On continuous multiplications on the two-sphere" Proc. Amer. Math. Soc., 7 (1956a), 942-947; "On a class of semigroups on E n" Proc. Amer. Math. Soc., 7 (1956b), 729-734; "On the structure of semigroups on a compact manifold with boundary" Ann. Math., 65 (1957), 117-143; "Semigroups with identity on a manifold" Trans. Amer. Math. Soc., 91 (1959), 380-389; "One-parameter semigroups in a semigroup" Trans. Amer. Math. Soc., 96 (1960), 510-517.

Mycielski, J., "Some properties of connected groups" Colloq. Math., 6 (1958), 162-166.

Nachbin, L., Topology and Order. D. Van Nostrant Co., Inc., Princeton, N. J., 1965.

Norris, E. M., "Inverse limits and embedding of compact totally disconnected machines" Math. Systems Theory, 5 (1971), 89-94.

Numakura, K., "On bicompact semigroups with zero" Bull. Yamagata Univ., 4 (1951), 405-412; "On bicompact semigroups" Math. J. Okayama Univ., 1 (1952), 99-108; "Theorems on compact totally disconnected semigroups and lattices" Proc. Amer. Math. Soc., 8 (1957a), 623-626; "Prime ideals and idempotents in compact semigroups" Duke Math. J., 24 (1957b), 671-679; "Naturally totally ordered compact semigroups" Duke Math. J., 25 (1958), 639-645.

O'Carroll, L., "Counterexamples in stable semigroups", Trans. Amer. Math. Soc., 147 (1969), 377-386.

Oberhoff, K. E., "Semilattice structures on trees" Semigroup Forum, 9 (1974), 54-73.

Paalman-De Miranda, A. B., Topological Semigroups. Mathematical Centre Tracts, 2nd Edition, Mathematiche Centrum Amsterdam, 1970.

Paterson, A. L. T., "Amenability and locally compact semigroups", Math. Scand. 42 (1978), 271-288.

Peter, F., and Weyl, H., "Die Vollstandigkeit der primitiven Darstellungen einer geschlussen kontinuierlichen Gruppe" Math. Ann., 97 (1927), 737-755.

Phillips, R. C., "Interval clans with non-degenerate kernel" Proc. Amer. Math. Soc., 14 (1963), 396-400.

Pontryagin, L. S., Topological Groups. Gordon and Breach, Science Publishers. Translated by Arlen Brown, 1966.

Precupanu, Anca, "Almost automorphic functions on topological semigroups", Rev. Roumaine Math. Pures Appl., 25 (1980), 783-788.

Pym, J. S., "Idempotent measures on semigroups" Pac. J. Math., 12 (1962), 682-698.

Reilly, N. R., "Transitive inverse semigroups on compact spaces" Semigroup Forum, 8 (1974), 184-187.

Rosen, W. G., "On invariant means over compact semigroups" Proc. Amer. Math. Soc., 7 (1956), 1076-1082.

Rothman, N. J., "Linearly quasi-ordered semigroups" Proc. Amer. Math. Soc., 13 (1962), 352-357; "On algebraically irreducible semigroups" Duke Math. J., 30 (1963), 511-517.

Rozet, A. I., "Topological semigroups with ideal chain conditions", Vesci. Akad. Navuk BSSR Ser. Fiz.-Mat., 1978, 11-16.

Ruppert, W., "On compact locally connected semigroups with identity" Semigroup Forum, 18 (1979a), 51-78; "On semigroup compactifications of topological groups" Proc. R. Irish Ac., 79 (1979b), 179-200; "The structure of semitopological monoids on compact connected manifolds" Math. Zeit., 170 (1980), 15-42.

Schützenberger, M. P., "D representation des demi-groupes", C. R. Acad. Sci. Paris, 224 (1957), 1194-1996.

Schwarz, S., "Remark on the theory of bicompact semigroups" Mat. Fyz. Casopis Slovensk. Acad., 5 (1955), 86-89.

Selden, J., "A note on compact semigroups" Notices Amer. Math. Soc., 8 (1961), 588.

Sheldon, W. L., Point Transitive Actions by Abelian D-Semigroups. Dissertation, Univ. of Mass., 1970.

Shershin, A. C., Introduction to Topological Semigroups. University Presses of Florida, Miami, 1979.

Sneperman, L. B., "Representations of topological semigroups by continuous transformations" Dokl. Akad. Nauk USSR, 167 (1966), 768-771; "Finite-dimensional representations of some bicompact topological semigroups" Dokl. Akad Nauk USSR, 176 (1967), 538-540; "On the theory of characters of locally bicompact topological semigroups" Math. USSR Sbornik, 6 (1968), 508-532; "The imbedding of topological semigroups in topological groups" Math. Zametki, 6 (1969), 401-409; "The representations of a locally bicompact semigroup into discrete ones" Dokl. Akad. Nauk USSR, 14 (1970), 1068-1070; "Weakly uniform monothetic semigroups" Semigroup Forum, 9 (1974), 275-277.

Stadtlander, D., "Thread actions" Duke Math. J., 35 (1968), 483-490; "Semigroup actions and dimension" Aqua. Math., 3 (1969), 1-14; "Actions with topologically restricted state spaces" Duke Math. J., 37 (1970), 199-206; "Actions with topologically restricted state spaces II" Duke Math. J., 37 (1972a), 199-206; "A structure theorem for d-classes of compact semigroup actions" Math. Systems Theory, 5 (1971b), 141-144; "Actions on plane continua" Math. Systems Theory, 7 (1974), 338-343.

Stepp, J. W., "A note on maximal locally compact semigroups" Proc. Amer. Math. Soc., 20 (1969a), 251-253; "D-semigroups" Proc. Amer. Math. Soc., 22 (1969b), 402-406; "Locally compact Clifford semigroups" Pac. J. Math., 34 (1970), 163-176; "Semilattices which are embeddable in a product of min intervals" Proc. Amer. Math. Soc., 28 (1971a), 81-86; "Semilattices which are embeddable in a product of min intervals" Semigroup Forum, 2 (1971b), 80-82; "The lattice of ideals of a compact semilattice" Semigroup Forum, 5 (1972), 176-180.

Storey, C. R., "The structure of threads" Pac. J. Math., 10 (1960), 1429-1445; "Threads without idempotents" Proc. Amer. Math. Soc., 12 (1961), 814-818.

Stralka, A. R., "The Green equivalences and dimension in compact semigroups" Math. Zeit., 109 (1969a), 169-176; "Rees-simple semigroups" J. London Math. Soc., 1 (1969b), 705-708; "The topological structure of a D-class" Bull. Austr. Math. Soc., 1 (1969c), 289-432; "Extending congruences on semigroups" Trans. Amer. Math. Soc., 166 (1972), 147-161; "The Green equivalences and dimension in compact monoids II" Semigroup Forum, 10 (1975), 8-24.

Tamura, T., "On compact one-idempotent semigroups" Kodai Math. Sem. Rep., 1954, 17-20.

Taylor, W., "Some constructions of compact algebras" Ann. Math. Logic, 3 (1971), 395-434.

Tanana, B. P., "Topological Semigroups whose subsemigroups are closed", Ural Gos. Univ. Mat. Zap, 10, (1977), 196-219.

Tymchatyn, E. D., "A one-parameter semigroup which meets many regular D-classes" Can. J. Math., 21 (1969), 735-739.

Ursell, J. H., "On two problems of Mostert and Shields" Proc. Amer. Math. Soc., 14 (1963), 633-635.

van der Waerden, B. L., Modern Algebra. Volume I. Frederick Ungar Pub. Co., New York, 1949.

Wallace, A. D., "Extensional invariance" Trans. Amer. Math. Soc., 70 (1951), 97-102; "A note on mobs" Acad. Bras. de Cienc., 24 (1952), 329-334; Notes on Topological Semigroups, 1953a (unpublished); "A note on mobs II" Acad. Bras. de Cienc., 25 (1953b), 334-336; "Cohomology, dimension, and mobs" Summa Bras. Math., 3 (1953c), 43-54; "Indecomposable semigroups" Math. J. Okayama U., 3 (1953d), 1-3; "Inverses in Euclidean mobs" Math. J. Okayama U., 3 (1953e), 23-28; "Topological invariance of ideals in mobs" Proc. Amer. Math. Soc., 5 (1954), 866-868; "On the structure of topological semigroups" Bull. Amer. Math. Soc., 61 (1955a), 95-112; "Structural ideals" Proc. Amer. Math. Soc., 6 (1955b), 634-638; "One-dimensional homogeneous clans are groups" Koninkl. Ned. Akad. van Wet.-Amsterdam, 17 (1955c), 578-580;

"The Rees Suschkewitch structure theorem for compact completely simple semigroups" Proc. Nat. Acad. Sci., 42 (1956a), 430-432; "The gebietstreue in semigroups" Koninkl. Ned. Akad. van Wet.-Amsterdam, 18 (1956b), 271-274; "Ideals in compact connected semigroups" Ned. Akad. Wetench. Proc. Soc., 18 (1956c), 535-589; "Retractions in semigroups" Pac. J. Math., 7 (1957); 1513-1517; "Remarks on affine semigroups" Bull. Amer. Math. Soc., 66 (1960), 110-112; "Acyclicity in compact connected semigroups" Fund. Math., 50 (1961a), 99-105; "Problems on semigroups" Colloq. Math., 8 (1961b), 223-224; "A theorem on acyclicity" Bull. Amer. Math. Soc., 67 (1961c), 123-124; "Relative invertability in semigroups" Czech. Math. J., 11 (1961d), 480-482; "Relative ideals in semigroups I" Colloq. Math., 9 (1962a), 55-61; "Problems concerning semigroups" Bull. Amer. Math. Soc., 68 (1962b), 447-448; "Relative ideals in semigroups II" Acta. Math., 14 (1963), 137-148; "The position of C-sets in semigroups" Proc. Amer. Math. Soc., 6 (1965), 639-642; "Recursions with semigroup state spaces" Rev. Roum. Math. Pures et Appl., 12 (1967), 1411-1415; "Externally induced operations" Math. Systems Theory, 3 (1969), 244-245.

Ward, L. E., Jr., "Mobs, trees, and fixed points" Proc. Amer. Math. Soc., 8 (1957), 798-804; "On a conjecture of R. J. Koch" Pac. J. Math., 15 (1965), 1429-1433.

Weyl, H., The Classical Groups. Princeton University Press, Princeton Math. Series, Princeton, N. J., 1946.

Whyburn, G. T., "Non-alternating transformations" Amer. J. Math., 56 (1934), 294-302.

Wong, James C., "Topological semigroups and representations", Trans. Amer. Math. Soc., 200 (1974), 89-109.

Williams, W. W., Admissibility of Semigroup Structures on Continua. Dissertation, Louisiana State Univ., 1969.

Wright, F. B., "Semigroups in compact groups" Proc. Amer. Math. Soc., 7 (1956), 309-311.

Yuan, John, "On the construction of one-parameter semigroups in topological semigroups", Pacific J. Math., 65 (1976), 285-292.

INDEX

T

U

W

Z